African
Mathematics

African Mathematics

History, Textbook and Classroom Lessons

BY

Robin Walker

&

John Matthews

REKLAW EDUCATION LIMITED
London (U.K.)

First published in 2014 by Reklaw Education

ISBN-13: 978-1500667399

ISBN-10: 1500667390

CONTENTS

OPENING REMARKS

African Mathematics: History, Textbook and Classroom Lessons aims to meet several objectives. Firstly, the book gives a historical overview of Africa and its contribution to Mathematics. Secondly, the book provides the teacher and the learner with study materials that can be used in the classroom. Thirdly, the book introduces some of the numerical patterns and puzzles that has fascinated one of the authors John Matthews. Fourthly, the book introduces the lectures, classes and workshops that the two authors teach on these subjects.

Mathematics has an interesting history in Africa. This history forms the first part of the book. The earliest known mathematical artefact in human history is the Lebombo Bone. Thought to be 37,000 years old, it was discovered by archaeologists in South Africa. Scholars believe that the number system carved into the bone represents a lunar calendar. Later mathematical evidence comes from the Ishango region of Central Africa, Ancient Egypt, Medieval North Africa, Ethiopia, Medieval West Africa and Medieval Central Africa.

Mathematics teachers have asked for material that can be used in the classroom where these African mathematical ideas can be used, tested or even challenged. To this aim, the second part of the book presents classroom material that can engage children. Each lesson demonstrates a mathematical principle followed by classroom exercises that the pupils can attempt. We have provided perhaps 21 hours worth of teaching and learning material.

The third part of this book was written wholly by John Matthews, a secondary school mathematics teacher. It contains a sample of his tips and short cuts that any pupil of mathematics can profitably use to improve their engagement with number.

Mr Matthews also believes that students of mathematics should enjoy detecting patterns in number sequences. In the fourth part of the book he presents a sample of his own discoveries in this area. He also presents classroom exercises that the pupils can attempt.

Finally, the book contains information on the classes, lectures and workshops that the authors deliver on mathematics education and also Black or African heritage teaching. We also give details of how to contact us.

Read and enjoy.

Robin Walker & John Matthews 2014

PART ONE

AN OUTLINE OF THE HISTORY OF AFRICAN MATHEMATICS

INTRODUCTION

In the News

Ibrahima Diallo Sambegou of Guinea solves 270-year-old math problem announced the web site Hob Nob Drive. *Great Mathematician in Guinea* announced the website News of the South. In fact numerous African websites have been buzzing with an interesting story since 2013 about a mathematician from the West African nation of Guinea (i.e. Conakry). Apparently, he devised a solution to a famous mathematical puzzle called Goldbach's Conjecture.

Guinean solves a math problem 270 years old

Figure 1. Internet news story of the brilliant West African mathematician, Ibrahima Diallo Sambegou.

In 1742 Russian mathematician, Christian Goldbach, sent a letter to his contemporary Leonhard Euler in which he surmised "every even number greater than two can be written as the sum of two primes." This was an interesting idea ... but was it true? The letter launched one of the oldest and best-known unsolved problems in number theory and in all of mathematics.

Many mathematicians have attempted to test Goldbach's Conjecture since the eighteenth century, but a Guinean researcher, Ibrahima Diallo Sambegou, worked on the problem over a 14 year period and arrived at a solution. Originally a journalist, he became a mathematician. Moreover he was working against time. American mathematicians were also racing to get there first.

"For illustration, we see that $6 = 3 + 3$, $8 = 3 + 5$, $3 + 7 = 10$ or $5 + 5$, $30 = 11 + 19 = 13 + 17$, $100 = 17 + 83$ ... is it true [of] any even number? This is the clue," Ibrahima said. The numbers that he mentions in building these equations--3, 5, 7, 11, 13, 17, 83--are all prime numbers, i.e. only divisible by themselves and by one to produce a whole number.

However, it is said that Ibrahima has been knocking on all doors to get his work validated. Finding little practical support in his own country, Guinea, Ibrahima decided to go to Dakar, Senegal, to validate his results at the mathematics institute there.

Africa and Mathematics

Which ever be the case, mathematics is an incredibly broad subject. Most scholars give the impression that unless one is dealing with pure abstract intellectualism, as Ibrahima was, then one is not doing real mathematics. Certainly African intellectual traditions, such as those of the Ancient Egyptians, the mediaeval Moors, and the mediaeval university traditions of Timbuktu and Katsina, were engaged in the kind of narrow intellectualism that we have come to associate with mathematics, but the field itself is much broader than this.

Professors Mamokgethi Setati and Abdul Karim Bangura, co authors of the superb *African Mathematics: From Bones to Computers*, cite several examples of the practical use of mathematics in diverse African cultures. Among these were:

> beautiful examples of geometrical mural decoration of a house front in Zinder (Niger), mural paintings from the Zande people in northeastern Congo, a Swahili plaster work design from the northern Kenyan coastal region, a Nubian stencil wall painting, a beautifully plaited strip design that decorates part of the

wall above the door of the house of a Bamileke chief in Cameroon, a wall decoration motif from Ghana, the roof structure of a Fulani house in Cameroon, a schematic plan of a large Massa farmhouse enclosure in Cameroon, and the circular settlement structure of Zulu cattle keepers in South Africa.

These examples show the practical applications to which Africans put mathematical ideas and techniques. It also opens our minds to bigger possibilities of what we think mathematics is.

Setati and Bangura cite other examples:

Wood carving is another activity that summons geometrical exploration. African wooden boxes, seats, headrests, doorposts, pilasters, canoes and boats, spears, drums, pestles, spoons, cups, masks, and combs often have symmetric shapes, in addition to often being decorated with geometric designs. Various other domains of African life that invite geometrical exploration, imagination and creativity include metalwork, tattooing and other forms of body decoration, hairstyles, string figure and other games.

Figure 2. Front cover of the excellent *African Mathematics: From Bones to Computers*, currently the best overall book on African mathematics history and a major inspiration for this work.

This part of the book is an overview of the known history of mathematical engagement in Africa. I have tried to focus on the achievements of indigenous Africans and for this reason I have little to say on the achievements of mathematicians of Ancient Greek or Mediaeval Arabian origin even if living in Africa. This is not to belittle their intellectual achievements but to make it clear that their achievements need to be discussed elsewhere. I have included the mathematics of the Moors. Moorish mathematics was certainly a part of the common heritage of Mediaeval Islam encompassing people of many countries and races. However, my focus is on the Moors who, at that time, were still largely but not exclusively a Black people.

This is a relatively new area of study and much more remains to be done. There are still enormous holes in our knowledge of the subject. I pay my intellectual debts to the scholars who have gone before in picking up piece after piece of information. Professors Mamokgethi Setati and Abdul Karim Bangura have produced the most comprehensive single text on the subject. Other great contributors are Professor Cheikh Anta Diop, Professor Charles Finch, Professor Beatrice Lumpkin, Professor Ron Eglash, Professor Paulus Gerdes and Professor Claudia Zaslavsky.

This part of the book is an overview of the research that has already been done. The story begins 37,000 years ago in prehistoric Africa with the Lebombo Bone. Then we have the well-documented contributions of the Ancient Egyptians to many branches of mathematics. The Ancient Egyptians in the days before the Arabs and the Greeks got there were indigenous Africans. The story then turns to mathematics in Ethiopia, mathematics among the Moors, and finally mathematics in mediaeval West and Central Africa. Finally, I suggest new areas of research that scholars may wish address to advance our knowledge further.

Robin Walker

CHAPTER 1: AFRICAN PROTO-MATHEMATICS

Introduction

Mathematical thinking, numbers and shapes has had a long and interesting history in Africa. The earliest examples of mathematical objects in human history include the Lebombo and Ishango Bones found in Southern and Central Africa. What are these objects? What do they tell us? It must be understood that these artefacts are pre-historic. They are much older than civilisation as we know it, older than governments, older than organised states and before the advent of sophisticated technology, as we know it.

The Lebombo Bone

How did humans in prehistoric times keep track of different days? An answer to this question emerged when excavations near Border Cave in the Lebombo Mountains between South Africa and Swaziland led to the recovery of a small piece of a baboon fibula. Dating from approximately 35,000 BC, and 7.7 cm long, this artefact was found to have been inscribed with 29 clearly defined notches.

Though 37,000 years old, some writers claim it resembles the calendar sticks still used by the San people of modern Namibia. Some think it represents a lunar phase counter suggesting that the proto-mathematicians who inscribed the notches were women using it to keep track on menstrual cycles. A menstrual cycle is approximately the same length as a lunar cycle. Other writers suggest that the bone demonstrated the existence of a refined accounting system that helped early humans to grasp the concept of time. The general importance cannot be overstated: The Lebombo Bone represents the first clear evidence of calculation in human history.

Franz Gnaedinger, who has a website on very ancient calendars, points out that the 29 notches can easily be read as a lunar calendar. There are 30 spaces between and next to the 29 notches. One should read the spaces and notches as follows: 30 spaces plus 29 notches plus 30 spaces plus 29 notches plus 30 spaces, etcetera, yields 30 29 30 29 30 ... nights. Moreover, the information could just as easily yield 30 59 89 118 148 177 207 236 266 295 325 354 nights for 1 2 3 4 5 6 7 8 9 10 11 12 lunations.

Figure 3. The Lebombo Bone showing the 29 notches clearly incised. 35,000 BC.

How did Gnaedinger arrive at these numbers? A lunation is the average time of one lunar phase cycle. The length of any one lunar month can vary from 29.26 to 29.8 days. Most writers present it as approximately 29.5 days. Thus the Bone allows us to approximately calculate one lunation as 30 nights, two lunations as 59 nights (i.e. 30 + 29), three lunations as 89 nights (i.e. 30 + 29 + 30), four lunations as 118 nights (30 + 29 + 30 + 29), etcetera.

The Ishango Bone

The Ishango bone is a tool handle with notches carved into it. Jean de Heinzelin, a Belgian archaeologist of the Royal Institute of the Natural Sciences, unearthed it in the late 1950s. He excavated in the Ishango region of Zaïre (now called Congo) near Lake Edward. The bone tool was originally thought to have been over 8,000 years old, but a more sensitive recent dating has given dates of 25,000 years old.

CNN reported the evidence with great enthusiasm. Their reporter, Patricia Kelly, said the following about the find:

> Best described as a prehistoric calculator, it's a piece of animal bone just ten centimetres long--about four inches ... When the notches are counted, a series of number sequences emerges. But they suggest a number system based on ten, another based on twelve, as well as the knowledge of multiplication, and of prime numbers ... It's thought that this piece of quartz at the tip may have been used for writing or engraving. The Ishango Bone may also be proof that a highly advanced civilization existed in Central Africa fifteen thousand years before the emergence of Egyptian culture.

On the tool are 3 rows of notches, two of which add up to sixty. The number patterns represented by the notches are of great interest to scholars.

Row 1 shows three notches carved next to six, four carved next to eight, ten carved next to two fives and finally a seven. According to some authorities, the 3 and 6, 4 and 8, and 10 and 5, represent the process of doubling or 2*n*. However, CNN interpret this column very differently. They see the 3 and 6 as 9 (i.e. 12 - 3), the 4 and 8 as 12 (i.e. 4 + 8), the 10 and 5 as 15 (i.e. 12 + 3), and finally, the 5 and 7 as 12 (i.e. 5 + 7). Consequently they conclude it is evidence of a number system based on 12.

Row 2 shows eleven notches carved next to twenty-one notches, and nineteen notches carved next to nine notches. This is thought to represent 10 + 1, 20 + 1, 20 - 1 and 10 - 1.

Finally, Row 3 shows eleven notches, thirteen notches, seventeen notches and nineteen notches. 11, 13, 17 and 19 are the prime numbers between 10 and 20.

Professor Ron Eglash, an authority on African mathematical thought, points out that the doubling system used in the Ishango Bone is fundamental to many African counting and mathematical systems of more recent times. He demonstrates that it is common in African languages to have words for even numbers that mean 2*n* or "n plus n". For instance in the Shambaa language spoken in Tanzania, "8" is *ne na ne*, literally "four and four". Another writer pointed out that doubling was employed in multiplication and division techniques in West Africa. Finally authorities on Ancient Egypt have noted the persistent use of powers of two in Ancient Egyptian mathematics. Finally, another important authority provides evidence that Ancient Egypt's use of base-2 calculations derived from the use of base-2 in Sub-Saharan Africa.

Returning to the Ishango Bone, colleagues have pointed out to me that the claim of prime numbers is a very far stretch. This is a very abstract concept that may not have existed at this early time since it is based on additional ideas such as division.

Row one adds up to forty-eight (3 + 6 + 4 + 8 + 10 + 5 + 5 + 7 = 48). Rows two and three add up to sixty (Row 2 consists of 11 + 21 + 19 + 9 = 60, and Row 3 consists of 11 + 13 + 17 + 19 = 60). Forty eight and sixty are multiples of twelve. These numbers provide further evidence of a number system based on twelve. However, the centrality of numbers ten and twenty for the calculations in Row 2 and Row 3, also suggest a number system based on ten, the basis of the decimal system of counting. For example, on a modern decimal ruler 10 millimetres is 1 centimetre, and 10 decimetres is 1 metre.

Finally, microscopic investigations illustrate more markings on the bone of different indentations, shapes and sizes. This led Alexander Marshack, another investigator, to suggest that the Ishango bone was a lunar phase counter. Marshack thinks there is a close connection between different

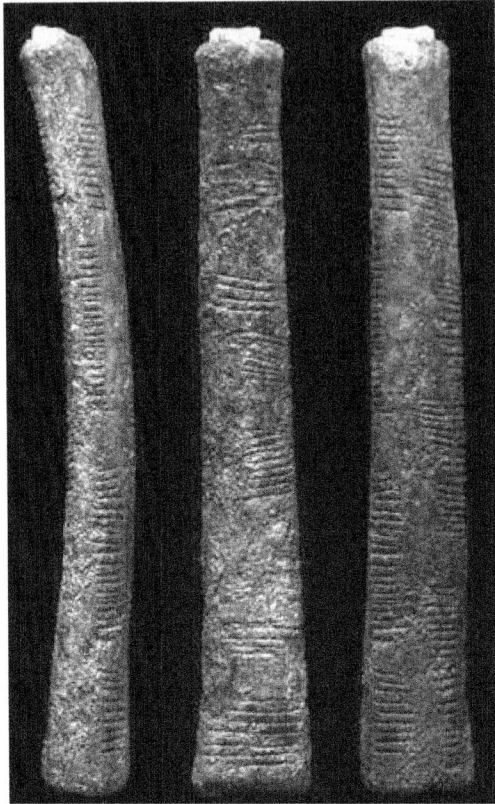

Figure 4. Three views of the Ishango Bone showing the different number patterns inscribed. 23,000 BC.

phases of the moon and the sequential notation contained on the bone, once the additional markings--hard to detect without a microscope--were taken into consideration. Moreover, he is of the opinion that the different markings of various sizes and shapes may originally have been a calendar of events of rituals and ceremonies.

I am of the opinion that the material on the bone may have multiple functions and so Marshack's views do not necessarily contradict the interpretations given by other writers and CNN. Consequently, the early mathematician(s?) behind the Ishango Bone may have some understanding of doubling or $2n$, addition, subtraction, multiplication, base 10, base 12, and may even have had some notion of prime numbers. Moreover the bone may even have been used to keep track of a calendar.

I shall give the final word on the Ishango Bone to Patricia Kelly of CNN who said: "It's thought Ishango Man's numbers system may have spread north following the River Nile into Egypt as well as into West Africa."

CHAPTER 2: ANCIENT EGYPTIAN MATHEMATICS

Historical Introduction

Ancient Egypt was the second major kingdom to emerge in African history after The Kingdom of Ta-Seti in Nubia. I controversially date the first Egyptian king, Narmer of the First ruling Dynasty as between 5660 BC and 5598 BC sticking closely to Egyptian records. Unlike the Kingdom of Ta-Seti, Egypt left behind a wealth of documentation and monuments allowing us to assess its mathematical contributions.

Egypt's first golden age was the Old Kingdom Period when the first six dynasties ruled (5660-4188 BC). Its second golden age was the Middle Kingdom Period when the eleventh and twelfth dynasties ruled (3448-3182 BC). During the Old and Middle Kingdom Periods, Egypt was unambiguously a Black African nation. Following a long period of alien occupation, indigenous Egyptians ruled a third time creating the New Kingdom Period when the eighteenth to the twentieth dynasties ruled (1709-1090 BC). Wealth, art and construction flourished during each of the golden ages leading to mathematical endeavour. However, Egypt after this period was conquered, first by other African nations of Libya and Kush, and then by Asiatics and Europeans who have dominated North Africa ever since. The Arabians who dominate North Africa today have been there since 639 AD. They have transformed North Africa both racially and culturally to what it is today. In this book, however, we are interested in the mathematical achievements of the indigenous Egyptians when Egypt was a Black African nation.

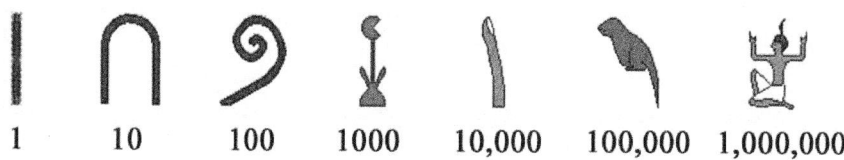

| 1 | 10 | 100 | 1000 | 10,000 | 100,000 | 1,000,000 |

Figure 5. The Ancient Egyptians had different hieroglyphic symbols for 1, 10 and the powers of ten.

Introduction to Ancient Egyptian Mathematics

Aristotle, the ancient Greek scholar, wrote: "And thus Egypt was the cradle of the mathematical arts." Plato, another ancient Greek scholar, reported that Socrates had heard that the Egyptian deity Thoth invented arithmetic, computation, geometry and astronomy. Proclus and Herodotus reported that the great Greek scholar Thales imported geometry from Egypt.

Our sources of knowledge concerning Ancient Egyptian mathematics comes from surviving documentary evidence. The most important documentary source is the Ahmose Papyrus, named after the scribe who copied it. Other scholars call this exquisite document the Rhind Mathematical Papyrus, after the Scottish collector who acquired it in 1858 and donated it to the British Museum. A second major source is the Moscow Papyrus written during the Middle Kingdom Period. It was taken to Russia in the mid twentieth century, ending up in the Moscow Museum of Fine Arts. These two texts contain a total of 112 mathematical problems and their solutions.

Other sources include a mace head from the time of Pharaoh Narmer of the First Dynasty which contains his war booty including 120,000 prisoners, 400,000 oxen, and 1,422,000 goats. The Egyptian Leather Roll is a table of 26 decompositions into unit fractions dating to the Middle Kingdom Period. The Berlin Mathematical Papyrus contains two problems of simultaneous equations. The Reisner Papyri are four rolls in very poor condition that record volume calculations relating to temples. Finally, the Kahun Papyrus contains six scattered mathematical fragments, not all of which have been deciphered. Like the other mathematical papyri, they too date from the Middle Kingdom Period. Which ever be the case, one fact remains inescapable, the mace head and the papyri indicate that a large proportion of what we today consider high school mathematics was created by the Ancient Egyptians.

Moreover, scholars have investigated Egyptian units of measure and the mathematical relationships between them. In addition scholars have investigated the mathematical relationships encoded in the proportions of Ancient Egyptian monuments particularly the Great Pyramid of Giza. This research has revealed Egyptian ideas of irrational numbers.

The Egyptians had a decimal system with special symbols for the numbers 1, 10, 100, 1,000, 10,000, 100,000 and 1,000,000. They even had a concept that translates as "millions of millions of years" which is regarded as representing infinity.

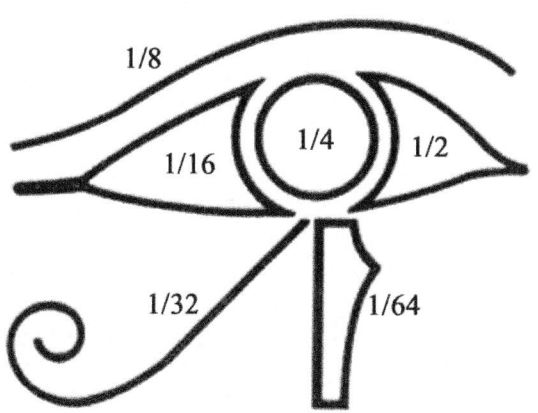

Figure 6. The Horus Eye Fractions. This was one of the ways the Ancient Egyptians presented fractions. When added together, all of these fractions total 63/64.

They had fractions, but these were always given as 1/2, 2/3, 1/3, 1/4, 1/5, 1/6, etcetera. With the exception of 2/3, all the other fractions were expressed as one over two, one over three, and so on. These are called unit fractions. If the Ancient Egyptians wanted to express 3/4, they would write 1/2 + 1/4. In a similar vein, if they wanted to express 2/5, they would write 1/3 + 1/15. As clumsy and difficult as this system seems to us moderns, it remains a fact that the Ancient Greeks adopted it. Moreover, the Europeans continued to use it right up to the beginning of the Renaissance period.

The Egyptians were adept at handling fractions, however. Problem 33 of the Rhind Mathematical Papyrus, for example, required the student to calculate 16 + 1/56 + 1/679 + 1/776 + 10 + 2/3 + 1/84 + 1/1358 + 1/4074 + 1/1164 + 8 + 1/112 + 1/1358 + 1/1552 + 2 + 1/4 + 1/28 + 1/392 + 1/4753 + 1/5432 = 37!

Mathematical Proof

The Egyptians had a notion of proof. Problem 67 of the Rhind Mathematical Papyrus reads as follows: "Behold, this shepherd comes to the livestock census bringing 70 bulls. The livestock steward says to the shepherd: 'What a small number of bulls you bring! Where, pray, are the many animals that you should have?' The shepherd answers him: 'What I bring you is 2/3 of 1/3 of the herd you entrusted to me. By my reckoning, I find that I have the whole herd.'"

The document then says: "Correct procedure: You multiply 70 by 4 1/2; The result is 315. That was the number entrusted: 315. 2/3 of which is 210. 1/3 of which is 105. 2/3 of 1/3 of which, is 70. That is what he brought."

The last line above begins "[t]hat is what", or more generally, "this is it" or *"na pu"* in the ancient language of the Egyptians. Professor Obenga, one of our best authorities on Ancient Egypt, says that "this is it" is the equivalent of QED that we use today. QED stands for "quod erat demonstrandum" in Latin and means "that which was to be proved." This is important evidence that shows that the Egyptians thought logically and used "correct procedure" to arrive at a conclusion or *"na pu"* the equivalent of QED.

Problem 58 ends with a final section of the problem where the scribe goes over to check the answer, step-by-step. Next to this is the word *"seshemet"* which means "proof" or "reflection."

How did the Ancient Egyptians handle multiplication?

When multiplying two numbers together, the Egyptians would make one number the multiplier and the other number the multiplicand. They would arrange the numbers into two columns respectively:

Multiplier   Multiplicand

Suppose they wanted to multiply five by seventeen, five would be the multiplier and seventeen would be the multiplicand. Always starting from number one, they would keep doubling the number until the number five was reached. They would also double the multiplicand starting with the number seventeen. The results would look like this.

Multiplier	Multiplicand
1	17
2	34
4	68

Since they were multiplying by five, the Egyptians would use the information concerning the multiplier of 1 (and the corresponding 17) and the multiplier of 4 (and the corresponding 68). They chose these figures because 1 + 4 = 5. Therefore the calculation would be 17 + 68 = 85. They would disregard the 2 and the corresponding 34.

Multiplier	Multiplicand
1 Use this	17 Use

2 Reject this	34 Reject this
4 Use this	68 Use
1 + 4 = 5	17 + 68 = 85

Suppose they wanted to multiply six by seventeen, six would be the multiplier and seventeen would be the multiplicand. Always starting from number one, they would keep doubling the number until the number six was reached. They would also double the multiplicand starting with the number seventeen. The results would look like this - just like before.

Multiplier	Multiplicand
1	17
2	34
4	68

Since they were multiplying by six, the Egyptians would use the information concerning the multiplier of 2 (and the corresponding 34) and the multiplier of 4 (and the corresponding 68). They chose these figures because 2 + 4 = 6. Therefore the calculation would be 34 + 68 = 102. They would disregard the 1 and the corresponding 17.

Multiplier	Multiplicand
1 Reject this	17 Reject this
2 Use this	34 Use
4 Use this	68 Use
2 + 4 = 6	34 + 68 = 102

To give another example, suppose they wanted to multiply fifteen by nine, fifteen would be the multiplier and nine would be the multiplicand. Always starting from number one, they would keep doubling the number until the number fifteen was reached. They would also double the multiplicand starting with the number nine. The results would look like this.

Multiplier	Multiplicand
1	9
2	18
4	36
8	72

Since they were multiplying by fifteen, the Egyptians would use the information concerning the multiplier of 1 (and the corresponding 9), the multiplier of 2 (and the corresponding 18), the multiplier of 4 (and the corresponding 36) and the multiplier of 8 (and the corresponding 72). They chose these figures because $1 + 2 + 4 + 8 = 15$. Therefore the calculation would be $9 + 18 + 36 + 72 = 135$.

Multiplier	Multiplicand
1 Use	9 Use
2 Use	18 Use
4 Use	36 Use
8 Use	72 Use
$1 + 2 + 4 + 8 = 15$	$9 + 18 + 36 + 72 = 135$

Suppose instead they chose 9 to be the multiplier and 15 to be the multiplicand, the end result would be the same but the methodology would look like this.

Multiplier	Multiplicand
1 Use	15 Use
2 Reject	30 Reject
4 Reject	60 Reject
8 Use	120 Use
$1 + 8 = 9$	$15 + 120 = 135$

Since they were multiplying by nine, the Egyptians would use the information concerning the multiplier of 1 (and the corresponding 15), and the multiplier of 8 (and the corresponding 120). They chose these figures because $1 + 8 = 9$. Therefore the calculation would be $15 + 120 = 135$. They would disregard the 2 (and the corresponding 30) and the 4 (and the corresponding 60).

Professor Charles Finch cites Richard Gillings, an authority on Egyptian mathematics, as writing the following:

> These additions were made easier for the scribe by virtue of a special property of the series 1, 2, 4, 8, 16, 32 ... for any integer can be uniquely expressed as the sum of some of its items ... We do not know whether or not the scribes were explicitly aware of this but they certainly used it, just as do the designers of a modern electronic computer, and this is surely a somewhat sobering thought.

How did the Ancient Egyptians handle division?

The methodology for handling division required the opposite process. Suppose the Egyptians wanted to divide 425 by 18 they would still arrange the numbers into two columns as before. The results might look like this:

1	18
2	36
4	72
8	144
16	288
32	576

Since 576 is larger than 425 the Egyptians would use the number 16 (and the corresponding 288). Now from 425 subtract the partial products, starting from the number 288. The results would look like this:

```
  425
- 288        (16 x 18)
  137
-  72        (4 x 18)
   65
-  36        (2 x 18)
   29
-  18        (1 x 18)
   11        remainder
```

The answer is $16 + 4 + 2 + 1 = 23$. The remainder is 11.

CHAPTER 3: THE RHIND MATHEMATICAL PAPYRUS

Introduction

The Rhind Mathematical Papyrus is a key source for the understanding of Ancient Egyptian mathematics. A. Henry Rhind, a Scottish lawyer, bought the scroll in 1858 from Luxor in Egypt. Egyptologists have found it convenient to name the scroll after the Scottish lawyer but the scroll itself contains enough information to give it a more accurate context. Originally composed during the time of the Middle Kingdom ruler, Pharaoh Amenemhet III (3242-3195 BC) of Dynasty XII, the Rhind Mathematical Papyrus is actually a copy produced many years later by a scribe called Ahmes or Iâhmesu. Ahmes or Iâhmesu made the surviving copy during the 33rd year of the Hyksos Pharaoh of Upper and Lower Egypt Aa-Wser-Râ. Aa-Wser-Râ belonged to Dynasty XV and could have reigned at any time between 2545 and *c.*1993 BC.

According to Professor Théophile Obenga's translation, the book begins with the following declaration: "Right method of investigating nature to know all that exists, all mysteries and all things secret. For that purpose, this papyrus roll was copied in the Year 33, fourth month of Akhet, under the Majesty of the King of Upper and Lower Egypt, Aa-Wser-Râ, May He live! And this was accurately copied from an ancient text produced in the time of the King of Upper and Lower Egypt, Ne-Maat-Ra, [b]y the scribe Iahmesu, who inscribed this copy."

Despite the broad claims for the document suggested by its opening declaration, the focus of the text is much narrower and focuses on mathematics. Mathematics was characterised as a set of principles or rules which, when systematically applied, reveals the secrets of nature and makes them intelligible.

This tells us that the writers of the original were producing a methodology and displaying a mentality similar to what we today call scientific method. The use of systematic methods moves us further and further away from received opinions and mass beliefs towards knowledge acquired in more objective ways. This is confirmed by other translations of the opening declaration. One writer says it means "Rules for enquiring into

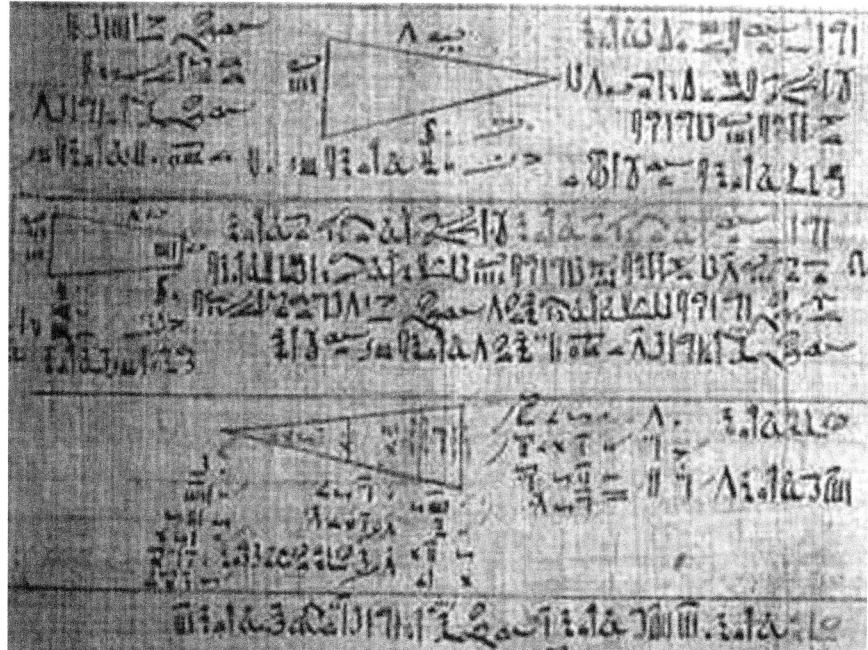

Figure 7. Detail from the Rhind Mathematical Papyrus, copied between 2545 and 1993 BC. Of supreme importance, this document contains a sizable proportion of what we moderns regard as high school mathematics.

nature." Another gives it as: "Rules for conducting research into nature. Yet another says: "[C]orrect method for enquiring into nature." As we saw in the last chapter, some of the problems end with *"na pu"* which means "this is it."

The text itself is a handbook that covers five broad themes. The opening section concerns arithmetic. The second section concerns stereometry, i.e. three dimensional measurement. The third section is on geometry. The fourth section contains calculations related to pyramids. The final section contains miscellaneous mathematical problems.

The scroll contains 87 mathematical problems, exercises or models. Problems 1 to 6 concern division. Problems 7 to 20 concern multiplication. Problems 21 to 23 concern subtraction. Problems 24 to 34 concern algebra. Problems 28 and 29 are 'think of a number' type of algebraic questions. Problems 35 to 38 are more first level equations. Problems 41 to 46 concern the volumes of shapes such as rectangular and cylindrical granaries. Problem 47 is the "Eye of Horus" system of fractions. Problems 48 to 55 concern the area of shapes such as the circle, rectangle, triangle and the trapezium. Problems 56 to 60 concern trigonometry. Problems 61 to 68 concern differential fractions. Problems 69 to 78 concern inverse proportions. Finally, Problems 79 onwards concern various miscellaneous issues.

Algebra

Problem 24 was the first of 11 simple algebraic problems in the Rhind Mathematical Papyrus. The question asks: "A quantity plus its $1/7^{th}$ becomes 19. What is the quantity?" The unknown number is called *'aha'* in the Ancient Egyptian language and means 'bundle, heap or lot'. In this context it refers to 'figure, quantity or number.' Today we would write *'aha'* as 'x.'

This is clearly an algebraic question asking the student to solve for one unknown. In modern algebraic terms the question looks like this:
$$x + x/7 = 19$$

The solution was given as:

$x + x/7 = 19$	
$8x/7 = 19$	This assumes the 'false position' that x is 7, thus $7 + 7/7 = 8$
$8x = 133$	Both sides have been multiplied by 7
$x = 133/8$	Both sides have been divided by 8
$x = 16\ 5/8$	This is the unknown quantity or *'aha'*
$x/7 = 2\ 3/8$	

Thus the question with the correct answer will now look like this: $16\ 5/8 + (16\ 5/8 \div 2\ 3/8) = 19$.

Problem 26 asks the question: "What number, a quarter of which, added to it, makes 15?" In modern algebraic terms the question looks like this:
$$x + x/4 = 15$$

The solution is given as:

$x + x/4 = 15$	
$5x/4 = 15$	This assumes the 'false position' that x is 4, thus $4 + 4/4 = 5$
$5x = 60$	Both sides have been multiplied by 4
$x = 60/5$	Both sides have been divided by 5
$x = 12$	This is the unknown quantity or *'aha'*

Thus the question with the correct answer will now look like this: 12 + 12/4 = 15.

The 'false position' method was in use for many centuries after the Egyptians pioneered the method. It remained in use right up to the Renaissance period.

Problem 28 is an algebra question of a different type. Scholars regard it as the first 'Think of a Number' problem ever recorded. The question reads as follows: "[T]ake a number; to it add 2/3 of it. Then from this sum subtract 1/3 of it; the remainder is 10. What is this number?"

The answer is nine. However, the calculation implied the following formula where x = 9:

$$x + 2x/3 - 1/3(x + 2x/3) = 10$$

Calculating Volume

Problems 41, 42 and 43 concern finding the volumes of cylindrical shaped granaries. The scribe calculated the volume of each granary or cylinder using a formula exactly equivalent to the modern formula of V = area of a circle x h, where V is the volume of the cylinder and h is the height of the cylinder. The actual formula the scribe used was $V = (8d/9)^2$ x h which is exactly equivalent.

For example, Problem 41 asks: "Find the volume of a cylindrical granary of diameter 9 and height 10."

The solution is given as

9 - 1/9 = 8	This is 8d/9 in the formula above
8 x 8 = 64	This is the answer squared or $(8d/9)^2$
64 x 10 = 640	This is the answer times height or $(8d/9)^2$ x h

Taking 1/9 away from the diameter is the equivalent of multiplying the diameter by 8/9. Multiplying the answer by itself is the same as squaring the answer. This implies that the formula for the area of a circle is $(8d/9)^2$ where d is the diameter of the circle.

The area of a circle is today thought of as $A = \pi$ x $(d/2)^2$ or $A = \pi$ x $d^2/4$ or even $A = \pi$ x r^2, where A is the area, d is the diameter, and r is the radius. Thus $A/(d/2)^2 = \pi$.

Crunching the numbers used in Problem 41 then $64/(9/2)^2 = \pi$. Simplifying, this gives a figure for π of 64/20.25. Other writers prefer to

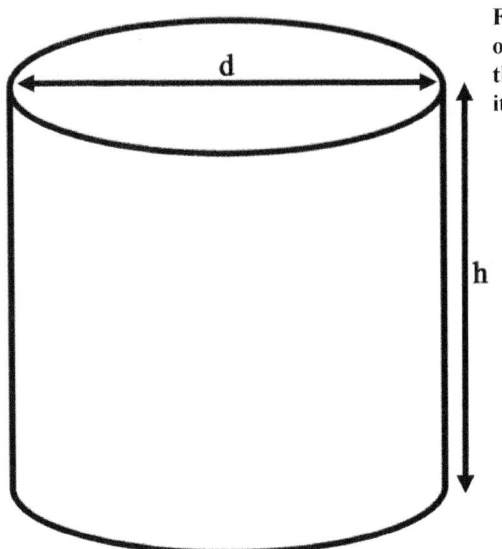

Figure 8. The RMP uses the formula of V = (8d/9)² x h for the volume of the cylinder. One could also present it as V = π x r² x h = πr²h.

present the next step of the equation as 64/(81/4) by squaring the top and bottom of the denominator. Multiplying the numerator and the denominator of the equation times 4, π becomes 256/81. Calculating 64/20.25 or 256/81 approximately equals 3.16049 to five decimal places. While this deviates from the rounded-up modern value for π = 3.14159 to 5 decimal places by 0.6%, this is the earliest documented implied value for π in a written document. As we shall see, there is an even earlier implied value for π built into the Great Pyramid of Giza.

Thus the volume of the cylinder could be presented as V = (8d/9)² x h or as V = (256/81) x (d/2)² x h or even V = (256/81) x (d²/4) x h. One could also present it as V = π x r² x h = πr²h.

Problem 44 involves cubing a number. This is where a number (in this case 10) is multiplied by itself and then multiplied by itself a second time. In measurement, a number 'to the power of one' or 10¹ has not been multiplied by anything and represents length. In this example, the length is 10 units long. In measurement, a number 'to the power of two' or 10² meaning multiplied by itself represents the area of a square. In this example, the area is 100 units squared. Finally, in measurement, a number 'to the power of three' or 10³ represents the volume of a cube. In this example, the volume is 1000 units cubed. Thus 10³ is 10 x 10 x 10 or as a formula V = a x a x a, where V is the volume and a is the length of each side of the cube.

Problem 45 requires extracting a cube root. This is just the opposite calculation where the volume of a cube has been given but the problem requires the scribe to find the length of a single edge of the cube.

Calculating Area

Problem 49 (and for that matter Problem 6 of the Moscow Mathematical Papyrus) concerns the area of a rectangle. Both papyri imply a formula of A = h x b where A is the area, h is the height of the rectangle, and b is the base of the rectangle.

Problem 50 addresses the area of a circle. The question reads as follows: "Example of a round field of diameter 9 khet; what is the area?"

The solution is given as
9 - 1/9 = 8   This is 8d/9 in the formula above, see Problem 41
8 x 8 = 64   This is the answer squared or $(8d/9)^2$
Its area is 64 setat.

Problem 51 (and for that matter Problem 4 of the Moscow Mathematical Papyrus) concerns the area of a triangle. The question reads as follows: "A

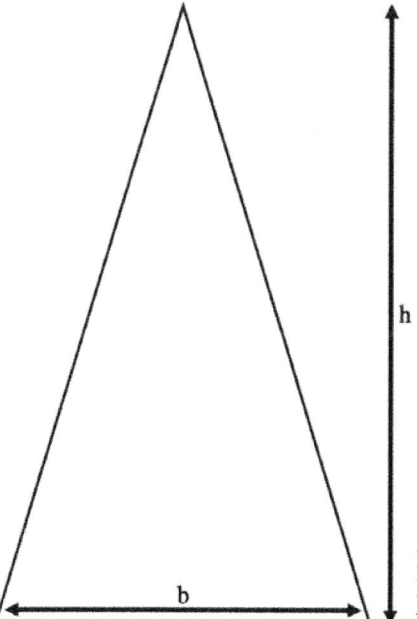

Figure 9. Both the RMP (and the MMP) use the formula A = 1/2 x b x h for the area of a triangle.

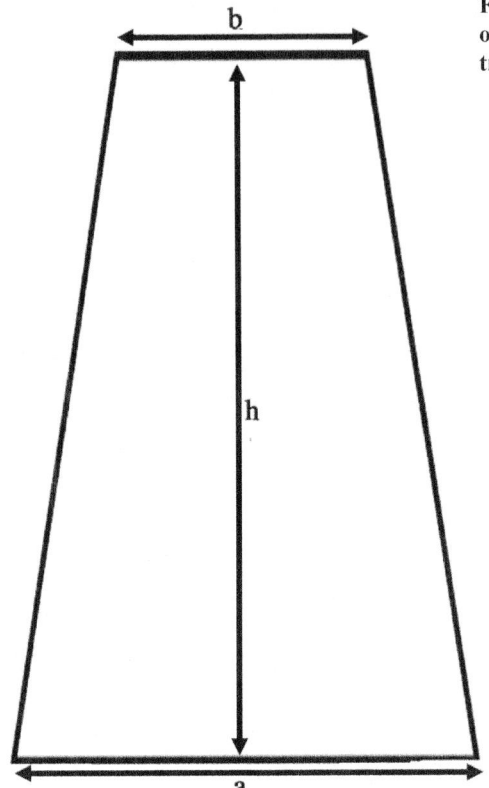

Figure 10. The RMP uses the formula of A = (a + b)/2 x h for the area of a trapezium.

demonstration of the calculation of a triangular plane. If asked: A triangle 10 rods high, 4 at its base; what's its area?" Both papyri, RMP and MMP, imply a formula of A = 1/2 x b x h where A is the area, h is the height of the triangle and b is the base of the triangle.

The solution is given as

4/2 = 2	This is the base halved (or 1/2 x b)
2 x 10 = 20	Base halved multiplied by the height (or 1/2 b x h) gives the area

Problem 52 concerns the area of a trapezium. The question asks: "[T]rapezium of 20 khet, with a large base of 6 khet and a small base of 4 khet. What is its surface? This implies a formula of A = (a + b)/2 x h where A is the area, a is the longer base of the trapezium, b is the shorter base of the trapezium, and h is the height of the trapezium.

The solution is given as

6 + 4 = 10	This is the longer base (or a) added to the shorter base (or b)
10/2 = 5	This is the average of the two base lengths (a + b)/2
5 x 20 = 100	Average base multiplied by the height (a + b)/2 x h gives the area

Trigonometry

Problem 56 (also 57 to 60) concerns trigonometry. The question reads as follows: "If a pyramid is 250 cubits high and the side of its base 360 cubits long, what is its seked?" The seked tells us the slope or gradient of the triangular faces of the pyramid. In trigonometry, the seked is the cotangent of the angle of the slope of the faces of the pyramid.

The scribe calculated 1/2 of the base of 360 cubits, which is 180 cubits. He has thus cut the pyramid in half from the pinnacle to the ground. Looked at from the side, the pyramid has now become two right angled triangles each with a base of 180 and a vertical height of 250. He then divided 180 by 250 to get 0.72 which is the cotangent of the slope. Thinking as a right

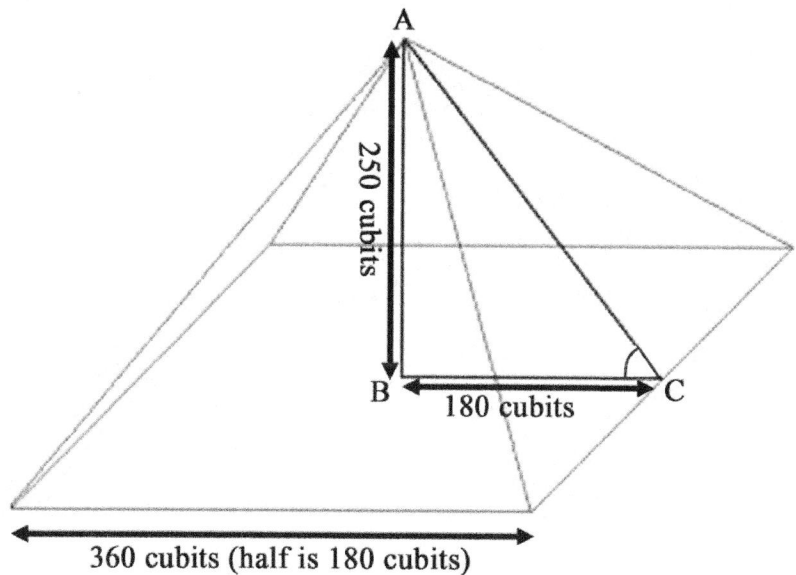

Figure 11. The contangent for the angle at C is calculated as BC/AB which in this case is 180/250.

angle triangle, this is the equivalent of dividing the length of the adjacent side (the base, i.e. 180) by the length of the opposite side (the height, i.e. 250). The scribe could have stopped here since this tells us that as we move 0.72 cubits along the base of the pyramid towards the centre, the face rises by 1 cubit in height.

However, the scribe went further and multiplied 0.72 by 7 to convert the units of the answer from cubits to another measure called hands or palms. One cubit equals 7 hands or palms. Thus 0.72 x 7 = 5.04 hands or palms. Thus as we move 5.04 palms along the base of the pyramid towards the centre, the face rises by 1 cubit in height.

What do ratios in cubits and palms mean to us moderns in degrees? A cotangent of 0.72 is equivalent to a tangent of 250 divided by 180 which gives 1.3888. Using a calculator to translate this into degrees, we get a slope of 54.24 degrees. This is more correctly presented as 54 degrees, 14 minutes and 46 seconds.

Mathematical Series

Problem 64 is an example of arithmetic progression. This is where each number in a series of numbers, is larger than the previous number by a fixed amount called the common difference. The question reads as follows: "Example of distributing difference. If it is said to thee, divide 10 hekats of barley for 10 men, the difference of each man to his neighbour 1/8 that is, [what is the share of each]?"

In modern terminology the question would be posed as: "The sum of 10 terms of an arithmetic progression is 10 and the common difference is 1/8. What are the terms of this series?" The mathematical scholar Richard Gillings shows that this involves the practical application of the formula S = n/2[2a + (n - 1)d] where S is the sum of terms, a is the first and lowest term, 1 is the last and highest term, d is the common difference, and n is the number of terms.

The calculation is given as

10/10 = 1	Average share of 1 hekat each or using the formula S/n
10 - 1 = 9	Number of differences is one less than number of terms or (n - 1)
1/8 x 1/2 = 1/16	Half of the common difference or d/2
9 x 1/16 = 9/16	d/2 x (n - 1)

1 + 9/16 = 1 9/16   This is the highest term or S/n + d/2 x (n - 1)
1 9/16 - 1/8 = 1 7/16 This is the second term
1 7/16 - 1/8 = 1 5/16 This is the third term
1 5/16 - 1/8 = 1 3/16 This is the fourth term
1 3/16 - 1/8 = 1 1/16 This is the fifth term
Then comes 15/16, 13/16, 11/16, 9/16 and 7/16 which is the last and lowest term term.

Problem 79 says:
Inventory of an estate:
Houses: 7
Cats: 49
Mice: 343
Barley seeds: 2401
Bushels: 16807
Total: 19607

According to Professor Obenga the text means: "Suppose that on an estate of 7 houses, each house had 7 cats, each cat killed 7 mice, each mouse ate 7 barley seeds, and each barley seed would have yielded 7 bushels; how many bushels would that make all told?"

This is a clearly a seven step geometric progression of $7 \times 7 \times 7 \times 7 \times 7$ = 16807. A geometric progression is a sequence in which each term succeeding the first is the product of the preceding term and a fixed number called the common ratio. The sequence is 7, 49, 343, 2401 and 16807. The common ratio is 7. If written today, Problem 79 would have been posed as: "Find the sum of 5 terms of the Geometric Progression whose first term is 7 and whose common ratio is 7." This also suggests a formula of $a_{n+} = a_n \times r$, where a_1 is the first term, r is the common ratio, and n = 1, 2, 3, 4, etcetera.

Problem 79 is also a problem with no immediate practical purpose showing that some Egyptian mathematics was purely theoretical or even just for fun. The other questions that we have examined so far all have some practical purpose.

Incidentally in mediaeval times the Italian mathematician Fibonacci wrote a mathematics problem with a similar idea. "Seven old women went to Rome: each woman had seven mules; each mule carried seven sacks: each sack contained seven loaves; and with each loaf were seven knives; each knife was put up in seven sheaths."

This Fibonacci problem would appear to have influenced the popular Mother Goose rhyme: "As I was going to St. Ives, I met a man with seven wives. Each wife had seven sacks, each sack had seven cats, each cat had seven kits. Kits, cats, sacks, wives, how many were going to St. Ives?"

Before leaving this problem, Professor Finch points out that the popularity of the number 7 in the Egyptian example is directly connected to the unusual method that the Egyptians used to do multiplication. For example, suppose we were told the following: "Suppose that on an estate of 7 houses, each house had 7 cats, how many cats would that make all told?" The Egyptians would have calculated it thus:

Multiplier	Multiplicand
1 Use	7 Use
2 Use	14 Use
4 Use	28 Use
$1 + 2 + 4 = 7$	$7 + 14 + 28 = 49$

Thus the number 7 means that all of the information in the calculation is used and none of it is rejected.

CHAPTER 4: OTHER EGYPTIAN PAPYRI

The Moscow Mathematical Papyrus

The Moscow Mathematical Papyrus was written during the Middle Kingdom Period. Housed in Moscow's Pushkin State Museum of Fine Arts, the papyrus was purchased by Vladimir Golenishchev sometime in the 1890s. The document contains approximately 25 mathematical problems, including how to calculate the length of a ship's rudder, the surface area of a hemispherical basket, the volume of a truncated pyramid, and various methods of solving for unknowns.

As we have seen, problems 4 and 6 deal with the areas of shapes (a triangle and a rectangle respectively). However, problems 10 and 14 are widely regarded as high points in the development of Ancient Egyptian mathematics, proving its sophisticated character.

Problem 10 asks: "Method of calculating a basket. If it is said to thee, a basket with an opening of 4 1/2 in its containing, Let me know its surface." The term basket is generally understood to be hemisphere, thus the problem concerns the difficult problem of calculating the surface area of a hemisphere. The Egyptian method produced an accurate answer of 32 which, again, assumes a figure for π as 64/20.25 or 256/81.

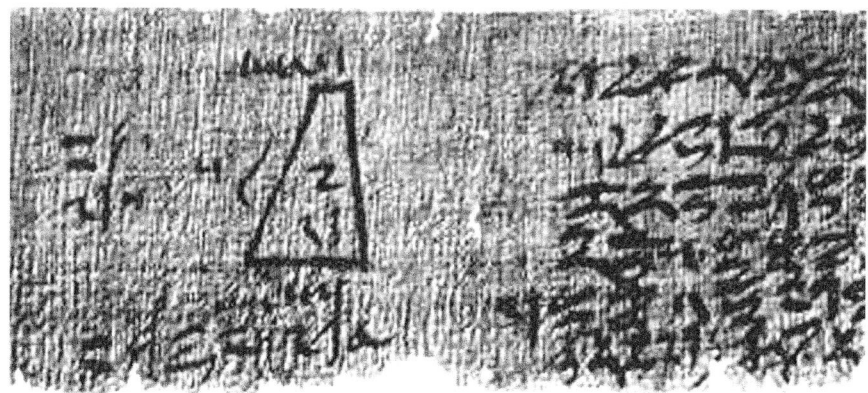

Figure 12. Detail from the Moscow Mathematical Papyrus, Problem 14.

The actual solution was given as:

4 1/2 x 2 = 9   This is the diameter multiplied by two or 2d

9 - 1/9(9) = 8   This is equivalent to 2d x (8/9)

8 - 1/9(8) = 7 1/9   This is equivalent to 2d x (8/9) x (8/9)

7 1/9 x 4 1/2 = 32   This is equivalent to 2d x (8/9) x (8/9) x d

The scribe added: "There results 32. This is its area. You have done it correctly."

Thus the formula the scribe used was S = 2d x (8/9) x (8/9) x d. This is also 2 x 64/81 x d^2, where S is the surface and d is the diameter. One could also present this as S = 2 x 64/81 x $(2r)^2$ or S = 2 x 4 x 64/81 x r^2 or even S = 2 x 256/81 x r^2 where r is the radius. Since 256/81 is the implied approximation for π in the RMP, then we could present the surface of a hemisphere as S = 2 x π x r^2 or $2\pi r^2$, the same way this formula is presented today.

Problem 14 asks: "Method of calculating a truncated pyramid. If it is said to thee, a truncated pyramid of 6 ellen in height, of 4 ellen of the base, by 2 of the top ..." Thus the problem involves the complex calculation of the volume of a frustrum (or a pyramid with the top cut off). The Egyptian method produced an accurate answer of 56. Clearly the scribe made a calculation exactly equivalent to the modern formula of V = h/3 (a^2 + ab + b^2), where V is the volume of the frustrum, a is the base, b is the length of the top, and h is the height.

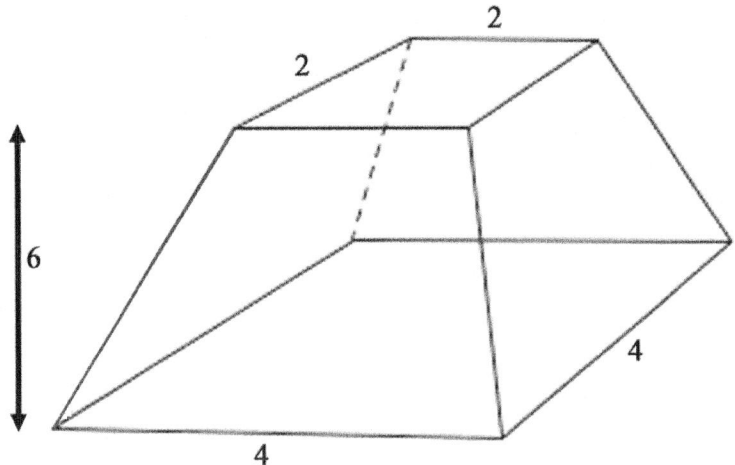

Figure 13. Frustrum with the measurements given in Problem 14.

The actual solution was given as:

$4^2 = 16$	This is a^2 in the formula above
$4 \times 2 = 8$	This is the ab (or $a \times b$) in the formula
$2^2 = 4$	This is b^2 in the formula
$16 + 8 + 4 = 28$	This is the $a^2 + ab + b^2$ in the formula
$6/3 = 2$	This is the $h/3$ in the formula
$2 \times 28 = 56$	This is the $h/3 \times (a^2 + ab + b^2)$ in the formula

Several historians of science and mathematics claim that the Egyptian method has not been improved upon in 4000 years.

The Berlin Mathematical Papyrus

The Berlin Mathematical Papyrus Problem 1 asks the question: "How to divide 100 into two parts such that the square root of one part is 3/4 of the square root of the other part."

In modern algebraic expressions, it looks like this:

$x^2 + y^2 = 100$, where $y = 3/4x$

Some writers present it like this:

$x^2 + y^2 = 100$, where $4y - 3x = 0$

Either way, all authorities are agreed that this is a problem requiring the calculation of two unknowns from two simultaneous equations.

How did they solve it?

The Egyptian scribe assumed the 'false position' that $x = 1$ and therefore y is 3/4 of that figure.

The first thing the scribe did was to square both sides and add them together:

For x this is $1^2 = 1$ and for y this is $(3/4)^2 = 9/16$

Added together this is $1^2 + (3/4)^2 = 1 + 9/16 = 1\ 9/16$

Next the scribe finds the square root of the total, so the square root of ($1 + 9/16$) = $1\ 1/4$

Following this, the scribe finds the square root of the other side of the equation, so the square root of $100 = 10$

Next the scribe divides the 10 by $1\ 1/4$ to get the figure for $x = 8$

Finally, the scribe calculates 3/4 of 8 which gives the value for $y = 6$

Thus $x = 8$ and $y = 6$

Thus the question with the correct answer will now look like this: $8^2 + 6^2$ = 100 or 64 + 36 = 100, where 6 = (3/4 x 8). Other authorities would present the second part of the problem as (4 x 6) - (3 x 8) = 24 - 24 = 0

Problem 2, when written in modern notation, requires the calculation of: $x^2 + y^2 = 400$, where 4y - 3x = 0

The answer is x = 16 and y = 12

Thus the question with the correct answer will now look like this: $16^2 + 12^2 = 400$ or 256 + 144 = 100, where (4 x 12) - (3 x 16) = 0.

The Kahun Papyrus

The Kahun Papyrus Plate VIII has a mathematical problem on it that has baffled scholars. Professor Diop speculates that it is dealing with the volume of a sphere with a hemisphere of 8 units in diameter. Other writers holding a different opinion think it concerns the volume of a cylindrical silo of 8 units in diameter and 12 units it height.

CHAPTER 5: EGYPTIAN MATHEMATICAL EVIDENCE FROM NON-DOCUMENTARY SOURCES

Irrational Numbers

Apart from the information in the papyri, what else can we deduce about Ancient Egyptian mathematical knowledge?

Professor Beatrice Lumpkin, an authority on Early African mathematics, points out that there is a demonstrable relationship between an Egyptian cubit and another Egyptian unit of measure called a double remen. A double remen is equal to the square root of 2 cubits. This is particularly interesting because the square root of 2 equals 1.4142 etcetera, an irrational number. We have already seen that the Egyptians were quite comfortable with extracting square roots. Many scholars thus see the relationship between the cubit and the double remen as convincing evidence that the Ancient Egyptians were familiar with the concept of irrational numbers. This raises a question: Did the Egyptians know of other irrational numbers?

Robert Bauval and Graham Hancock, co authors of the impressive *Keeper of Genesis,* report a figure of 3023.16 feet for the base perimeter of the Great Pyramid of Giza. This is the most famous of Egypt's 90 pyramids and represents the highest achievement of the Egyptian Old Kingdom Period. They also report a figure of 481.3949 feet for the vertical height of this monument. Since the radius of a circle x 2 = the diameter of a circle, and the diameter x π = the circumference of a circle, therefore π = the circumference divided by the diameter of a circle. Crunching some figures, this means: 481.3949 x 2 = 962.7898. Therefore π = 3023.16/962.7898 = 3.14. Compare this figure with the more accurate modern figure of 3.14159 to 5 decimal places. Π is an irrational number. The decimal places keep going on without end.

Scholars have raised the question of why the π proportion in the Great Pyramid (3.14) is more exact than the later RMP figure (60/20.25 or 256/81 also 3.16049). Professor Charles Finch has put forward a sophisticated theory in his great work on ancient and mediaeval African science and technology *The Star of Deep Beginnings.* He believes that the Egyptians wanted to make practical relationships between different branches of

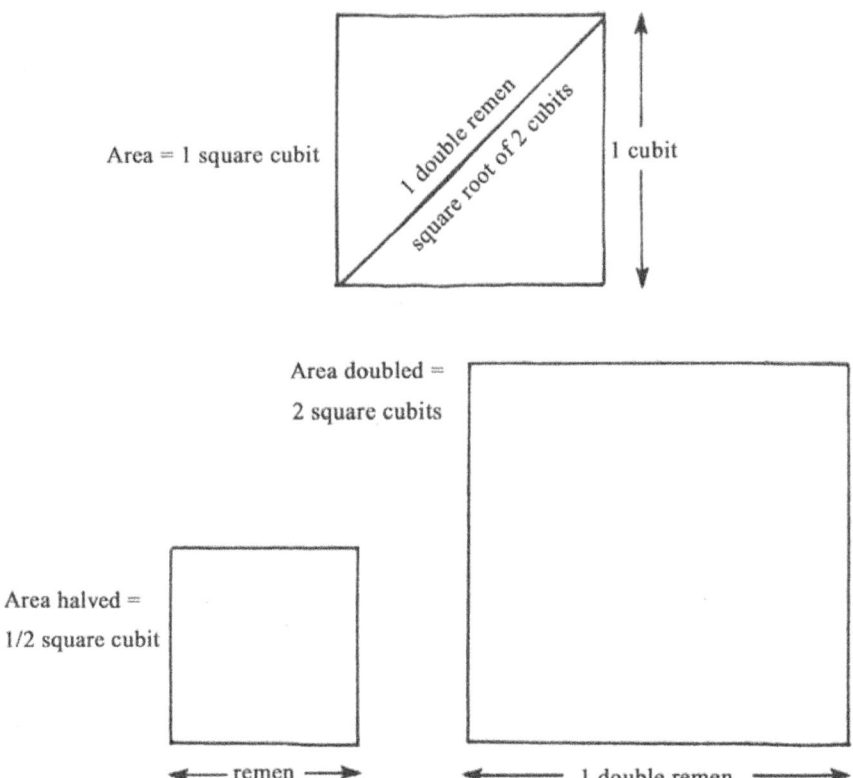

Figure 14. As Professor Beatrice Lumpkin points out with these squares, there is a demonstrable relationship between an Egyptian cubit and another Egyptian measurement called a double remen. A double remen is equal to the square root of 2 cubits. This is particularly interesting because the square root of 2 equals 1.4142 etcetera. It is an irrational number.

knowledge, in this case, mathematics and music. The mathematical relationship between the speed of the vibrations of the notes of a diatonic scale (or *do re mi fa so la te do*) are as follows from *do* to *re* the ratio of the speed of the vibrations is 8:9, from *re* to *mi* the ratio is 8:9, from *mi* to *fa* the ratio is 243:256, from *fa* to *so* the ratio is 8:9, from *so* to *la* the ratio is 8:9, from *la* to *te* the ratio is 8:9, but from *te* to *do* the ratio is 243:256. The significance of this is that 256/243 is EXACTLY one third of 256/81! Moreover, we have already seen how important the 8/9 ratio is in the Egyptian calculation of the circle. Thus the Egyptians, according to Finch, chose this value of π to be able to relate mathematics with music.

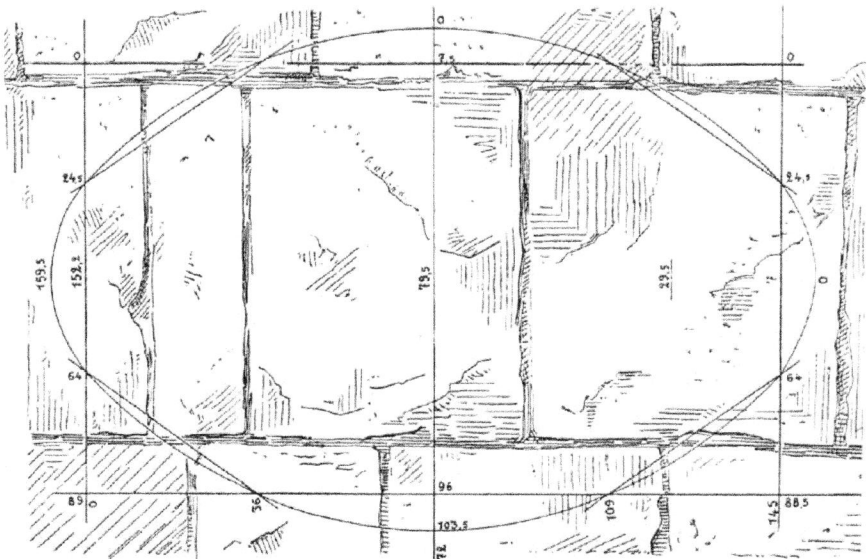

Figure 15. Ellipse from a Luxor Temple wall, *c.*1200 BC, as drawn by Ludwig Borchardt.

My own theory is much simpler and possibly less controversial than that of Finch. I regard the Great Pyramid of Giza as representing the pre eminent example of Old Kingdom Egyptian mathematical and scientific knowledge. I further believe that the RMP represents Middle Kingdom mathematical and scientific knowledge which was not quite as sophisticated. I believe that the First Intermediate Period that occurred between the Old and the Middle Kingdoms represents hundreds of years of political and cultural decadence that was not completely reversed when the Middle Kingdom Period arrived. Thus the inexact figure for π shows decadence.

Each of the four sides of the Great Pyramid is approximately 755.5 feet long against a vertical height if 481.3949 feet. This seems an arbitrary size to choose to build. However, the Egyptians actually reckoned the base as 440 cubits (half the base was 220 cubits), they reckoned the vertical height as 280 cubits, and the sloping height as 356 cubits. Thus the Egyptians thought of the Great Pyramid's dimensions in round numbers. These round numbers are not arbitrary.

Let us crunch some figures. If the sloping height of the pyramid is 356 cubits and half of the base is 220, then 356/220 = 1.618181. Compare 1.618181 with Φ = 1.618 to 3 decimal places.

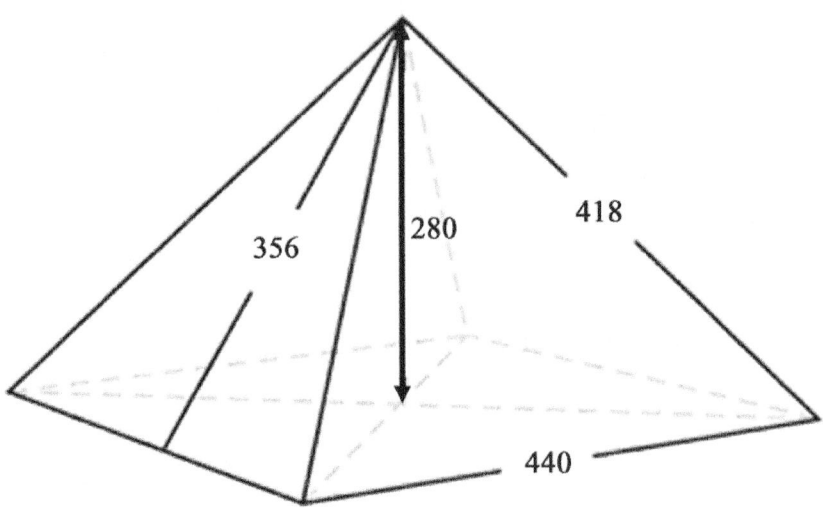

Figure 16. Dimensions of the Great Pyramid of Giza, *c.*4794 BC expressed in cubits.

This figure of Φ or *phi* is 1.618 to 3 decimal places, is also called the Golden Section (the actual figure, like π, is an irrational number of infinite decimal places). Snail's shells are based on this number. Artists and architects have found that if they base their designs on this number, the results are often aesthetically pleasing. It has even been claimed that a beautiful human face is also based on this figure!

Calculating the Area of an Ellipse

An elliptical oval was drawn on a wall of a Luxor Temple built during the time of Pharaoh Rameses III of the New Kingdom Period. It is intersected by a rectangle in such as way as to suggest that the rectangle was used to help calculate the area of the ellipse. In 1896 Ludwig Borchardt, a famous Egyptologist from Germany, published his discovery of this drawing. Scholars used to think that scholars in the Greek period were the first to attempt to calculate the area of this shape but clearly the Egyptians made an attempt hundreds of years before the Greek period.

The so-called "Pythagorean Theorem"

The Ancient Egyptians pioneered the 3, 4, 5 triangle which is the basis of the so-called 'Pythagorean Theorem.' They used this triangle to accurately calculate right angles. That the Egyptians knew this triangle, the following excerpt by Plutarch, the great Greco-Roman scholar, is most edifying:

> The Egyptians appeared to have figured out the world in the form of the most beautiful of triangles ... This triangle, the most beautiful of triangles, has its vertical side composed of three, its base of four, and its hypotenuse of five parts, and the square of the latter is equal to the sum of the squares of the two sides.

Moreover, Professor Finch notes that Sir Flinders Petrie, the great British archaeologist, found not one but several examples of Pythagorean triplets when he surveyed the Great Pyramid of Giza and the Pyramid field nearby.

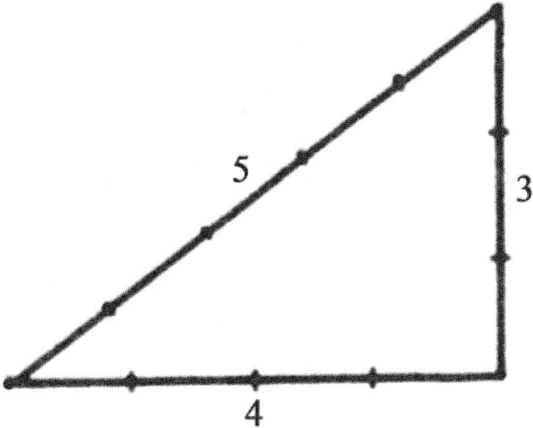

Figure 17. In some circles, particularly among European Americans, it has become customary to deny or challenge Egyptian knowledge of the 3 4 5 triangle and the theorem attributed to Pythagoras which is based on it, i.e. $3^2 + 4^2 = 5^2$. However, the Egyptian origin of this theorem is taught uncontroversially as fact in mainstream British schools. By knotting a rope with twelve knots equally spaced apart and tying the two ends of the rope together, the Egyptians used an ingenious method to calculate accurate right angles--necessary for their building projects. They did this by pulling the rope tight in different directions--particular the first, fourth and the ninth knot. Thus the rope will naturally take on a 3 4 5 triangle shape, giving an accurate right angle in one of the corners.

CHAPTER 6: MATHEMATICS IN THE TIME OF THE MOORS

Historical Introduction

The Moors were the early Black people from North West Africa. They created a great civilisation with splendid cities such as Marrakesh, Fez, Rabat and Meknes. Other Moors, in conjunction with the Arabs, marched into Spain in 710 AD and ruled the southern portion of Spain and Portugal until 1492. Four great dynasties held sway over the period of which two were unambiguously African, the Almoravid and the Almohades dynasties. The Moors (and Arabs) established great cities in Spain such as Cordova, Seville and Granada. Though the Moors were cleared out of Spain and Portugal between the fifteenth and the seventeenth centuries after losing control of the city of Granada, and have since lost power in North Africa to the Arabs, the legacy of the Moors in the evolution of mathematics cannot be denied or overlooked.

What do we know about Moorish mathematics?

The Science Museum in the Spanish city of Granada has excellent information on the science, technology and mathematics of the Moors and Arabs. Some of this section is based on the data from that museum. They, for example, point out that the exact sciences in former times were not as distinct and separate from each other as they are today. Thus the same scholar who penned a treatise on mathematics may also write learnedly on astronomy, physics, optics, or many other subjects.

Mathematical development was spurred on by practical concerns. Calculus and algebra were used to calculate the distribution of inheritances. Mathematics was used in architecture and in the surveying of land. Astronomers used mathematics to calculate times of Islamic prayer, the phases of the moon, and in the maintenance of the Islamic calendar.

Mathematical development was also spurred on by the movement of people through travelling. Pilgrims going to and from Mecca and other holy sites in the Islamic world bought books and met scholars from other parts of the Islamic world. Other scholars went on study trips to meet scholars and to learn from them.

Moorish scholars were familiar with the mathematical writings of early Greek mathematicians Apollonius, Archimedes, Euclid, Menelaus, Nicomano of Gerasus and Ptolemy. They were also familiar with the works of Arabic mathematicians Abu Kamil, Ibn Sinan and Tabit Ibn Qurra. Thus, it must be stressed the knowledge we discuss here was not exclusively African. The world of mediaeval Islam developed a scientific culture that was the common property of people from many different parts of the world.

Thus a mathematical work written by a Persian scholar in Baghdad would find its way into a school, madrassa or university in Africa or into Moorish and Arab-ruled Spain. Numerals originally invented by Hindu mathematicians were introduced into Europe by the Moors. The precursors to the nine numerals we use today were in use by at least the tenth century. Texts originally written by the Ancient Greeks were preserved, copied and commented upon by Islamic scholars of all colours. The Ancient Greek writings, themselves, show clear traces of the influences from Ancient Egyptian mathematicians. Finally, four of the Ancient 'Greek' writers may not have been Europeans after all. Some authorities claim that Euclid, Erathosthenes, Hypatia and Cladius Ptolemy were in reality indigenous Africans from Egypt or Libya that had European names.

Al-Hassar

How do you extract a square root from a number without using a calculator? Extracting square roots was one of a number of problems that occupied the mind of Abu Zakariya Muhammad ibn Abu Abd Allah Ayyash al-Hassar (meaning "the rush mat maker"). He was a 12th century Moorish scholar who once resided in the Moroccan city of Sebta. He was a mathematician, reader of the Qur'an, and a specialist in the division and distribution of inheritance. His *Book of Proof and Recall* is a mathematical handbook of calculation. It covered various topics (i) numbers, (ii) addition, subtraction, multiplication, and division of whole numbers and fractions, how to extract the square root of whole numbers or fractions, and (iii) summation of progressions of whole numbers and of their squares and cubes.

The book is of historical importance because fractions are written symbolically with the horizontal bar and the Hindu numerals--i.e. the ancestors of the digits that we use today. In the chapter on fractions he wrote: "but if you have to represent a fraction, then write the denominator under a [horizontal] line and above each its mentioned part." This is the first known use of the fraction bar like 1/2 or 2/7. A little later Fibonacci

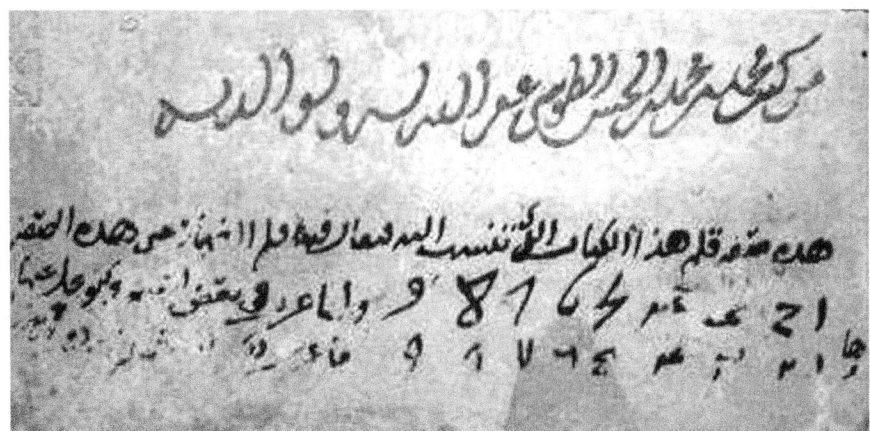

Figure 18. Detail from al-Hassar's *Book of Proof and Recall* showing at the bottom of the page both the Western and the Eastern Arabic forms of the nine Hindu numerals.

became the first European mathematician to employ it in his *Liber ab(b)aci* of 1202. One source goes further and speculates that al-Hassar's books "are probably the books that Fibonacci used when he learned math in North Africa, to bring back to Europe and teach people there."

His *The Complete Book on the Art of Number* contains 117 surviving folios. It was a key work of mathematics education from the time it was written up untill the sixteenth century. It dealt with whole numbers, decomposition of a number into prime factors, the common divisors and multiples, and the extraction of the cubic root of a whole number. Another part of the work, which has not been recovered, is thought to have concerned operations of fractions; the summation of the different categories of whole numbers; and an exposition of the algorithms that facilitate the calculation of various types of numbers.

Friedrich Katscher wrote an interesting article about Al-Hassar entitled: *Extracting Square Roots Made Easy: A Little Known Medieval Method - Al-Hassar's Description of the Method.* Al Hassar writes: "When it is said: which is the square root of 5, so take the next square number to 5, this is equal to 4, subtract it from 5, the remainder is 1, divide this by 4, that gives 1/4, and add this to the root of 4, which is equal to 2. That gives 2 1/4. And this is the approximate root of 5. Namely, when you multiply 2 1/4 [= 9/4] by itself you get [81/16 =] 5 1/16. The deviation is an excess of 1/16"

Al Hassar continues: "But if you want a closer approximation, then double 2 1/4, that gives 4 1/2. Divide 1/16 by 4 1/2. This gives 1/72 (an eighth of a ninth). Subtract this from 2 1/4 [or 2 18/72]. This leaves a remainder of 2 17/72 [or 161/72]. When you multiply this by itself you get [25921/5184 =] 5 1/5184, and this is nearer [to 5] than 5 1/16."

Finally: "If you want a closer approximation yet, double 2 17/72, divide $(1/72)^2$ [= 1/5184] by the result, and subtract what you get from 2 17/72. So the result will be an even closer approximation than the first and second roots. You may continue like this as far as you want."

Friedrich Katscher concludes:

> In other words, apply this method several times. Of course, in the process the numerator and the denominator of the fractions get longer and longer, and the multiplications become more and more unwieldy. Today's calculator or computer spares us the multidigit calculations that had to be done by hand in the past.

Ibn al-Yasamin

Ibn al-Yasamin al-Ishbilli (d. 1204) was a Moorish mathematician from Marrakech who was educated in the Spanish city of Seville where he was also a welcome visitor to the Almohad Dynasty court. He wrote on how to use the then new Indian number system in geometry and to calculate area. He is key to the transmission of Hindu numerals from India, then into Egypt, then Morocco (where he ran with it) and finally into Europe. These numerals have evolved into the so-called 'Arab' numerals used today.

Originally his reputation was built on a number of mathematical poems he wrote in the 1190s as teaching material. One was his *Poem on Algebra,* a work of 62 lines which explained the algebraic teachings of the great Islamic mathematician al-Khwarizmi. The success of the poem encouraged him to write another one that concerned the roots of irrational quadratic numbers. These are irrational numbers that are solutions to quadratic equations with rational coefficients. He wrote a third poem that summarised the algebraic method using the false position--a technique that dates back to the Ancient Egyptians (RMP Problem 24).

Of greater importance, qualitatively and quantitatively, was a work called *Fecundation of the Spirits with the Symbols of the Dust Ciphers.* The book was more than 200 folios in length and dealt equally with the sciences of calculation and contained chapters of geometry relative to the calculation of areas. Muslim scholars of the region tended to link these subjects together. Another reason for the importance of the book is its originality. Unusual for Maghrebian writers, he treats multiplication and division first before discussing addition and subtraction. His book had original things to say on algebra. In particular, he wrote about the abstract manipulation of polynomials (i.e. multiple terms). Finally the book is of key importance for the use and transmission of the 'dust ciphers' i.e. Hindu numerals into the 'western world.'

Muhyi l'din

Born in Spain, Muhyi l'din al-Maghribi (1220-1283) was an eminent Moorish astronomer. In the domain of mathematics, however, he is most famous for his work on trigonometry. He wrote *Book on the Theorem of Menelaus,* a mathematician from the Ancient Greek period, and *Treatise on the Calculation of Sines.*

In this second work he calculated an approximate value for the difficult problem of the sine of one degree achieved using two different methods. These calculations also led to him finding an approximate value for π.

He also considered the classical problem of doubling the cube. This means calculating the size of a new cube whose volume is exactly double the original cube. This is a complex problem requiring one to take the cube root of the number 2 as part of the calculation.

As well as commentaries and discussions of classic Greek works, he also wrote a particularly important discussion of a text that scholars generally call *Book XV of Euclid's Elements.* Although not actually written by Euclid, scholars have traditionally agreed to put his name on it. Which ever be the case, Muhyi l'din's commentary covers "the ratios of (1) the edges, (2) the

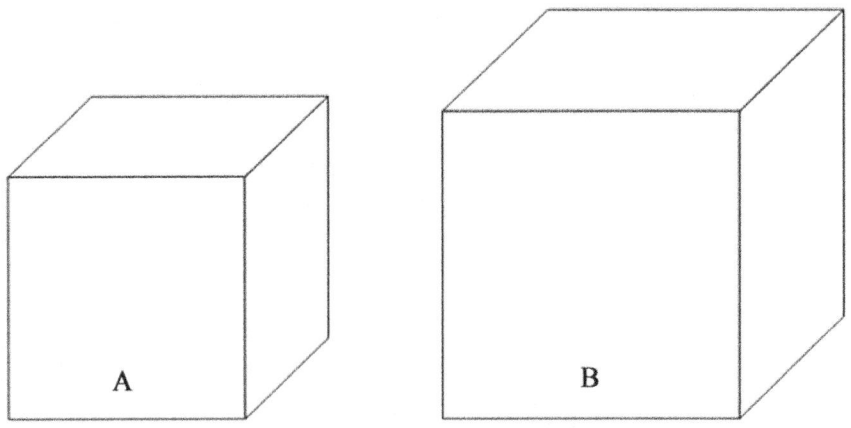

Figure 19. Two cubes where Cube B is twice the volume of Cube A.

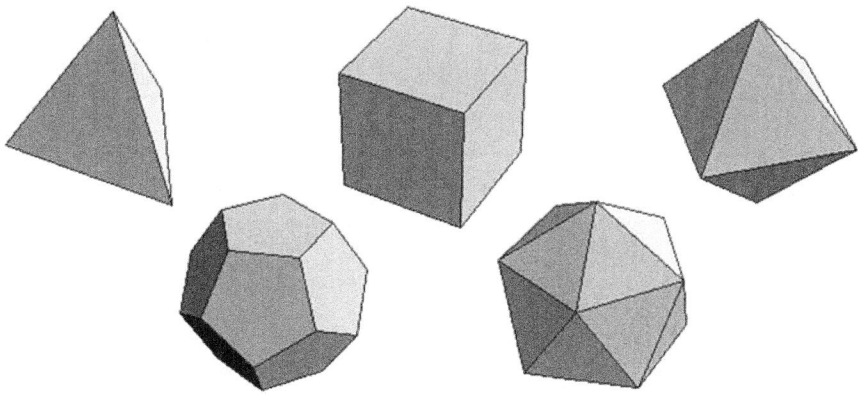

Figure 20. These are the five regular polyhedra.

faces, (3) the surface areas, (4) the perpendicular distances from the centre to a face and (5) the volumes of the five regular polyhedra inscribed in one sphere." Polyhedra are solid figures with many plane faces, typically more than six.

Al-Banna

Al-Marrakushi ibn al-Banna (1256-1321) was born in Marrakesh. He was one of a number of pioneering African mathematicians to use special symbols derived from Arabic letters that meant add, subtract, multiply, divide, unknown quantity, square, cube and equals. It was thus possible to write mathematical expressions with the same economy that we use today when we use the symbols +, -, x, ÷, x, x^2, x^3 and =.

Were these the first mathematicians ever to attempt this? Some claim that algebraic symbols may have been used in the Islamic East in the century before this.

Less controversially, al-Banna was the first mathematician ever to consider a fraction as a ratio between just two numbers. He is also the first scholar known to have used the expression 'almanac' (in Arabic al-manakh means 'weather') in a book containing astronomical and meteorological data.

As a pupil, he was schooled in geometry, Euclid's *Elements,* fractional numbers, and the impressive body of mathematical research that the Muslim world had contributed to mathematics over the preceding 400

Figure 21. 'Arabic' numerals as they appear in al-Banna's texts.

years. The Merinid Dynasty of Morocco encouraged a culture of learning and the city of Fez became their intellectual centre. At the university in Fez, Al-Banna taught all branches of mathematics including arithmetic, algebra, geometry and astronomy.

In his *The Lifting of the Veil,* among many other issues, he wrote on other numerical systems other than the system that we commonly use today, base 10. One source speculates that this research was "possibly influenced by their country's interaction with the Yoruba people to their south, who counted using base twenty." I believe this speculation is groundless but it is an interesting idea worth investigating.

The importance and prestige of Ibn al-Banna did not only come from his mathematical works. In fact, he distinguished himself from his Maghrebian predecessors by the richness and diversity of his production. Of the more than 100 titles of writings that are attributed to him, only 32 concern mathematics and astronomy. The others deal with disciplines that are very different, such as astrology, grammar, linguistics, logic and rhetoric.

His *Advertisement to the Intelligentsia* can be divided into two parts. The first part of the book contains the precise mathematical answers to questions that touch on varied domains of everyday life. Among these were the composition of medicaments, the calculation of the drop of irrigation canals, an arithmetical explanation of a verse of the Qur'an on inheritance, the determination of the hour of the third daily prayer, the explanation of frauds linked to instruments of measurement, the enumeration of delayed prayers which have to be said in a precise order, and the exact calculation of legal tax in the case of a delayed payment. The second part of the book has a collection of 17 little arithmetical problems presented in the form of poetic riddles.

Modern scholars have pointed out interesting mathematical ideas and results which appear in his books. Some are far too complex and difficult to present here, but two ideas are worth pointing out.

Firstly, he wrote on continued fractions and used them to compute approximate square roots. This appears to be an idea related to the writings of Al-Hassar mentioned earlier.

Secondly, another interesting idea is his results on how to sum a series. The method he used can be understood by the following formulae:

$$1^3 + 3^3 + 5^3 + ... + (2n - 1)^3 = n^2(2n^2 - 1)$$

Also:

$$1^2 + 3^2 + 5^2 + ... + (2n - 1)^2 = (2n + 1)2n(2n - 1)/6$$

CHAPTER 7: MATHEMATICS IN ETHIOPIA

Historical Introduction

Ethiopia has thousands of years of documented history. Starting from the city of Yeha before 500 BC, Ethiopia has had great periods of historical achievement centred on the cities of Axum, Lalibela, Gondar and Harar. The country was important as centres of trade, culture and religion-- Judaism, Christianity and Islam. The engineering skill behind the monuments in Axum and Lalibela baffles architects and engineers to this very day! Despite this, outside of the work of Professor Otto Neugebauer little seems to be known about its early mathematics.

The civilisations of Ancient Kush, Mediaeval Nubia and the Swahili Confederation were equally brilliant cultures. The monuments they left behind are as good as anything the rest of the ancient or mediaeval world had to offer. The interested reader should use Google Images to see monuments in Musawarat, Banganarti, Sogno Mnara and Kilwa for themselves as well as the evidence from the Ethiopian cities listed above. Again, despite all this evidence, little seems to be known about the early mathematics of these regions.

In this chapter, I shall present what is currently known about mathematics in Ethiopia.

An Unusual System of Multiplication

A few years ago, BBC 4 presented an interesting short film about multiplication in Ethiopia as part of a series about mathematics and number. Entitled *Go Forth and Multiply,* the film explained the millennia old system of multiplication used in Ethiopia by its traders and merchants. One of the products that the traders sold was coffee. Interestingly coffee was first consumed in Ethiopia and Jamaica now produces the finest coffee in the world.

If an Ethiopian trader wanted to multiply 11 and 15, he would put the numbers into two columns. He would place the 11 in one column and he would place the 15 in the other column.

In the first column he would continually halve the number ignoring the fractions. Thus 11 halved is 5 (i.e. ignoring the fractions), halved again is 2, and halved again is 1. In the other column he would double the numbers.

Thus 15 doubled is 30, doubled again is 60, and doubled again is 120. The two columns might look like this:

Halving column	Doubling column
11	15
5	30
2	60
1	120

There is a rule that one must IGNORE any even number(s) in the halving column AND the corresponding number(s) in the doubling column.

Consequently, we shall ignore the 2 and the 60. Our table now looks like this.

Halving column	Doubling column
11	15
5	30
1	120

Finally we add up the numbers in the doubling column to produce our answer which is 15 + 30 + 120 = 165.

The narrator commented that: "It seems unbelievable that a system can ignore fractions, even throw away parts of the calculation and still come up with the right answer."

The underlying principle of this system is doubling which, at its core, is base 2 arithmetic. The narrator explained the significance of this: "It's a system that seems completely foreign to Western eyes but in fact we use it thousands of times a day because it's this system that powers today's computers."

The attentive reader will have noticed the great similarity of this system with the one used in Ancient Egypt and the centrality of base 2 arithmetic dating back to the Ishango Bone.

Mathematical ideas in Ethiopian manuscripts

When is Easter? How do we calculate it? Ethiopia possesses interesting chronological ideas preserved in its mediaeval manuscripts. Professor Otto Neugebauer studied these manuscripts dating mostly from the fourteenth to the nineteenth centuries and wrote two books: *Ethiopic Astronomy and Computus* and *Chronography in Ethiopic Sources.*

As well as a hollow month of 29 days and a full month of 30 days, Neugebauer found indigenous Ethiopic terms in the manuscripts for zero, the hour, a solar cycle of 28 years, a jubilee of 49 years and a 'seal' of 76 years.

Neugebauer found that the Ethiopians had a lunar calendar. I note that it was much like the Lebombo Bone. The Ethiopians alternated hollow months of 29 days with full months of 30 days. This produced a lunar year of 354 days. They calculated this as $(29 + 30) \times 6 = 354$. Once every four years, they would add an extra day to the lunar year making 355 days.

The Ethiopians also had an Enoch year of 364 days. Coming from the astronomical information in one of their holy books, the *Book of Enoch,* they calculated this as four equal seasons of 91 days each i.e. $91 \times 4 = 364$. 91 days is also equal to 13 weeks of 7 days since $13 \times 7 = 91$.

There was also an Alexandrian year which averaged 365 1/4 days. For three out of every four years they would have a 365 day year named after the three evangelists in the Bible, Matthew, Mark and Luke. The year consisted of 12 months of 30 days each plus five extra days i.e. $(12 \times 30) + 5 = 365$. On each fourth year, named after the evangelist John, they would add a sixth extra day giving a total of 366 days. This is similar to the leap year calculation that we use today.

The difference between the Alexandrian year and the Lunar year for three out of four years was 11 days i.e. $365 - 354 = 11$ days. For the years named after the evangelist John, the difference was also 11 days i.e. $366 - 355 = 11$ days.

Most importantly, Neugebauer found that Ethiopian scholars and historians used a 532 year cycle as their basis for calculating time. What is the basis of the 532 year cycle?

When the Ethiopians compared the Alexandrian year to the Lunar year, they found that the lunar year was shorter by 11 days i.e. $365 - 354 = 11$. After two years it was 22 days shorter i.e. 11×2. After three years it was 33 days shorter i.e. 11×3. At this point the Ethiopians would add another lunar month of 30 days. Over a period of 19 Alexandrian years, they discovered the need to add a total of 7 lunar months. Since there are 12 lunar months in a year, there are 228 lunar months in a 19 year period i.e. $19 \times 12 = 228$. Since it was necessary to add 7 extra lunar months, there

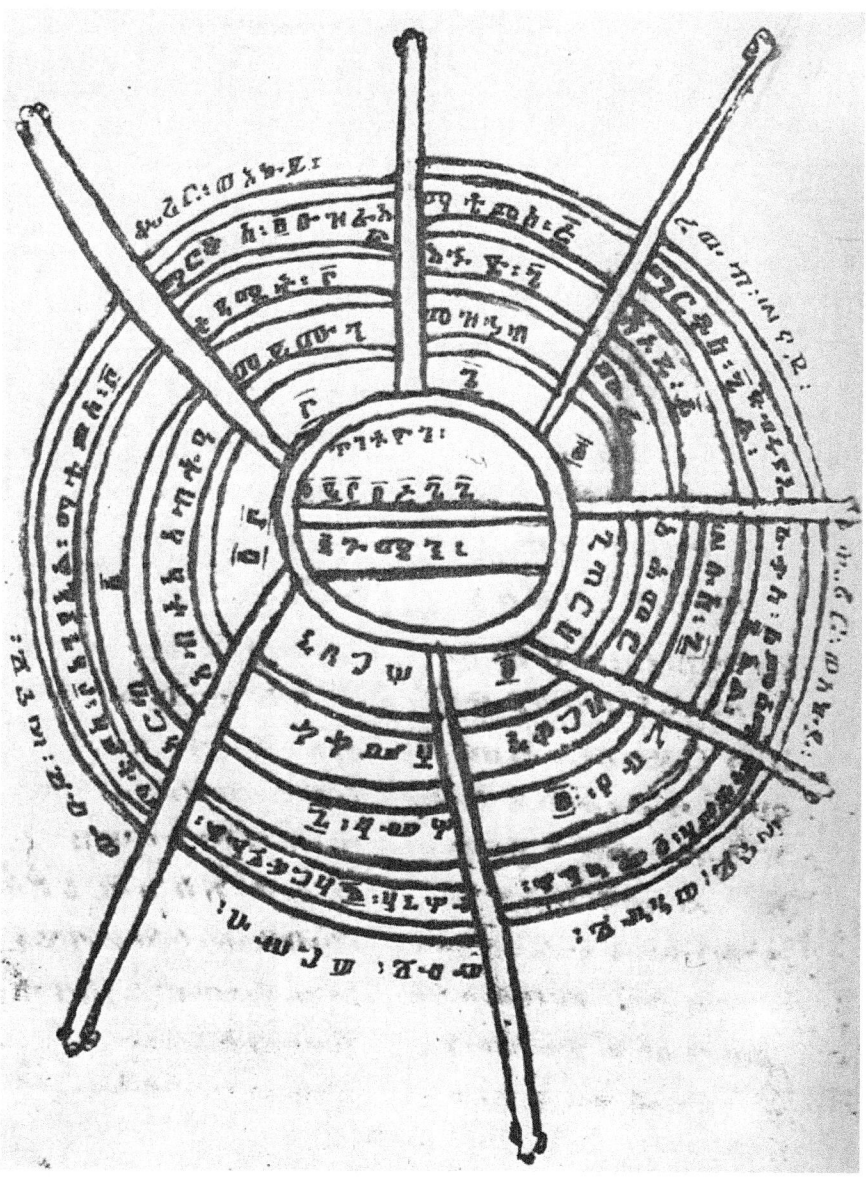

Figure 22. Diagram from an old Ethiopian manuscript that conceptually links together the seven day week and the four years of the evangelists.

were 235 lunar months in every 19 Alexandrian years. This 19 year cycle became a key part of Ethiopian calculations.

The Ethiopians had a seven day week. Sunday is called the Christian Sabbath, the names for Monday to Friday are based on the Ethiopian numbers 2 to 6, and Saturday is called the First Sabbath. With each passing year the days of the week at each date change. After a period of 7 years, the days of the week and the dates coincide.

Thus the 532 year cycle embodies the 19 year cycle, the 4 year cycles of the years of the evangelists, and the 7 day week. 532 is the lowest common multiple of 19, 4 and 7 since 19 x 4 x 7 = 532. It also embodies another Ethiopian unit of time called a 'seal' which is of 76 years. A seal is composed of 4 lots of 19 year cycles i.e. 4 x 19 = 76. There are 7 seals in 532 years since 7 x 76 = 532.

Modulo Arithmetic

How did the Ethiopians calculate Easter? They made extensive use of modulo arithmetic.

Modulo arithmetic is where there is a limit on the highest number that can be calculated from a procedure. Once that number has been reached, we return to one and complete the calculation. For instance, the highest number on an analogue clock is 12. After this the next number is 1. So if a question required the calculation of 11 pm + 3 hours, the answer would be 2 am and NOT 14 pm! Mathematicians would pose this same question as: What is 11 + 3 modulo 12? This tells us that 12 is the highest possible number. The next number after 12 is 1, then 2, etcetera. If the question was posed as: What is 11 + 23 modulo 12? The answer cannot be 34 since the highest possible number is 12. The quickest way to arrive at an answer is to divide 34 by 12 and the answer is the remainder. 34 divided by 12 is 2, the remainder is 10. Thus the answer to 11 + 23 modulo 12 is 10. When writing the formulae, it is customary to shorten modulo to 'mod'.

In explaining these ideas, I shall use two symbols that the reader may be unfamilar with (i) $\equiv$ and (ii) $\leq$. The first symbol means 'equivalent to'. It is similar to, but not quite the same as, 'equals'. Therefore '$c \equiv$ W mod 19' means that c is equivalent to W where 19 is the highest possible number. The second symbol means 'less than or equal to'. Therefore 'I 15 $\leq m \leq$ II 13' tells us that the lowest possible figure for m is month I day 15 and the highest possible figure for m is month II day 13.

The Ethiopians reckoned time in big blocks of data called Eras. The oldest era was the Era of the World. Beginning 5492 BC this represented the beginnings of creation from Ethiopia's Biblical perspective. Beginning 5036 BC was the Era of Bizan, named after an Ethiopian monastery. Beginning 8 AD was the Era of Incarnation. Beginning 284 AD was the Era of Martyrs. Finally beginning 360 AD was the Era of Grace. Ethiopian scholars believed that the Second Coming of Christ was to be in Year 7000 of the Era of the World. One of the manuscripts suggested that this means that human existence would total 2,548,000 days (i.e. 364 days x 7,000 years). Another manuscript suggested that this meant a total of 2,556,750 days (i.e. 365 1/4 x 7,000).

The Ethiopians saw time as repeated cycles of 532 years which they tabulated in 28 tables of 19 years each (i.e. 28 x 19 = 532). The ultimate aim of these tables was to help calculate the day and date of Easter each year since this is a moveable lunar and solar festival. In 2014, for example, Easter Sunday was celebrated on 20 April. In 2015 it will be celebrated on 5 April.

W	c	e	m	t	f
1	1	0	30	3	13
2	2	11	19	4	2
3	3	22	8	5	25
4	4	3	27	6	9
5	5	14	16	1	1
6	6	25	5	2	21
7	7	6	24	3	6
8	8	17	13	4	25
9	9	28	2	6	17
10	10	9	21	7	2
11	11	20	10	1	22
12	12	1	29	2	13
13	13	12	18	4	5
14	14	23	7	5	18
15	15	4	26	6	10
16	16	15	15	7	1
17	17	26	4	2	21
18	18	7	23	3	6
19	19	18	12	4	26

Figure 23. This is taken from the first of the 28 19-year tables showing information for years 1 to 19. W means years of the world, c means cycle number of years, e means epact, m means matqe'e, t means tentyon and f means fasika. What these terms mean and how they are calculated will be explained over the next few pages.

In calculating the day and date of Easter for any one of the 532 years, the Ethiopian scholars made a number of calculations using a number of interrelated formulae. I shall illustrate this by calculating Easter for Year 472 of their system.

First, they calculated a parameter that Professor Neugebauer calls 'cycle number of years, from 1 to 19.' Neugebauer symbolised it with a c. This parameter was calculated for any of the 532 years, symbolised by W meaning 'Years of the World', modulo 19. Thus 19 was the highest possible number from the calculation. The next number after this was 1, then 2, etcetera. Since Year 472 is much higher than 19, we need to divide it by 19 and the remainder is the answer. Thus:

$c \equiv W$ mod 19
$472/19 = 24$ remainder 16
Thus $c \equiv 16$ mod 19

Second, they calculated a parameter called the Epact. Symbolised in Neugebauer's analysis by an e, the Epact measured the running total of differences between the 354 day lunar year and the 365 day solar year. After one year the difference was 11 days. After two years the difference was 22 days. After three years the Ethiopians would add an extra lunar month. Consequently, the difference in the third year was reduced from 33 days, to just 3 days. To calculate the Epact for any given year, they used a formula equivalent to the one given below:

$e \equiv (c - 1) \times 11$ mod 19
$(16 - 1) \times 11 = 165 \equiv 15$ mod 19
Thus $e \equiv 15$ mod 19

Third, they used the Epact to calculate a third parameter called the Matqe'e. Symbolised by Neugebauer using an m, it was the date of the Jewish New Years Day. Ethiopia is home to a very ancient Black Jewish community. The earliest date that Matqe'e could fall on was day 15 of the first Ethiopian month (i.e. I). The latest date that it could fall on was day 13 of the second Ethiopian month (i.e. II). It was calculated by subtracting the Epact from 30. Thus:

$m = 30 - e$ where I $15 \leq m \leq$ II 13
$m = 30 - 15 =$ I 15 (i.e. month I, day 15)

Fourth, they used the Matqe'e to calculate Passover or p. The earliest date that Passover could fall on was day 25 of the seventh Ethiopian month (i.e.

VII). The latest date that it could fall on was day 23 of the eighth Ethiopian month (i.e. VIII). It was calculated by adding 190 to the Matqe'e, modulo 30. Thus:

$p = m + 190$ where VII $25 \leq p \leq$ VIII 23
$p = 15 + 190 = 205 \equiv 25$ mod 30
Thus $p =$ VII 25 (i.e. month VII, day 25)

Fifth, they calculated a parameter called the Tentyon or t. This was the day of the week on which the Ethiopian New Years Day fell (not to be confused with the Jewish New Years Day). It was the first day of the first month, (i.e. I 1) but there was a formula they used to work out which day of the week this was. By tradition, the Ethiopian scholars *who compiled these manuscripts* considered Wednesday to be the first day of the week, Thursday became the second day, etcetera. To calculate the Tentyon, they used a formula equivalent to the one given below:

$t \equiv W - 1 + 1/4 (W - 1)$ mod 7
$472 - 1 + 1/4 (472 - 1) = 588.75$
588.75 rounded up becomes 588 or $\equiv 1$ mod 7
t = Day 1 = Wednesday

In the next two steps, the Ethiopian scholars calculated the day of the week for Matqe'e (Jewish New Years Day) and the day of the week for Passover. Professor Neugebauer shows that these calculations did not require any complex formulae. They relied on simple arithmetic. For instance, if the first day of the first Ethiopian month (i.e. I 1) was a Wednesday, then the Matqe'e being on I 15 is also going to fall on a Wednesday. If the Passover was 190 days later and 189 days is a multiple of 7, then the 190th day will be a Thursday.

For the final calculation of Easter Sunday or f (meaning Fasika), the Ethiopians relied on simple arithmetic. Easter Sunday was always celebrated as the Sunday after Passover. Since Passover was on Thursday VII 25, then Easter Sunday was three days later. The earliest date that Easter Sunday could fall on was day 26 of the seventh Ethiopian month. The latest date that it could fall on was day 30 of the eighth Ethiopian month. Thus:

$f =$ VII $25 + 3 =$ VII 28   where VII $26 \leq f \leq$ VIII 30

The Ethiopians used other formulae to calculate other festivals of the year. Chief among these were the Beginning of Fast (or *bf*) and the Ninevah festival (or n). They calculated these by formulae equivalent to the ones given below:

$bf = f - 55$ mod 30   where VI $1 \leq bf \leq$ VII 5

$n = bf - 14$ mod 30

W	c	t	f
457	1	3	17
458	2	4	2
459	3	5	22
460	4	7	13
461	5	1	28
462	6	2	18
463	7	3	10
464	8	5	29
465	9	6	14
466	10	7	6
467	11	1	26
468	12	3	10
469	13	4	2
470	14	5	22
471	15	6	7
472	16	1	28
473	17	2	18
474	18	3	10
475	19	4	23

Figure 24. This is taken from one of the 28 19-year tables showing information for years 457 to 475. W means years of the world, c means cycle number of years, t means tentyon and f means fasika.

Using the example of Year 472, the Beginning of Fast was on VI 3 and the Ninevah festival was on V 19.

Using the Ninevah festival, the Ethiopians calculated other Christian festivals. Among these were Mount Olive (which was $n + 41$), Hosanna (which was $n + 62$), Synod ($n + 93$), Ascension ($n + 108$), Pentecost ($n + 118$), Prayer of Salvation ($n + 121$), and Fast of the Apostles ($n + 126$). Both the actual formulae and the modulo 30 versions appear in the Ethiopian manuscripts.

Base 60

Some manuscripts discuss sexagesimal or base 60 fractions of the day. Others contain a concept called kekros. What does this strange term mean? Professor Neugebauer mentions diagrams and texts from Ethiopian manuscripts that describe the changing illumination of the waxing or waning moon. The manuscripts present this as a linear sequence which varies between 4 kekros and 60 kekros in 15 steps of 4 kekros each. Thus each kekros represents 1/60 of the diameter of the illuminated fraction of the lunar disk. Other manuscripts show that 4 kekros means 48 minutes and 60 kekros means 12 hours. Thus 1 kekros means 12 minutes in this instance. A different manuscript shows that 15 kekros means 1 hour and 360 kekros means 24 hours. Bringing this evidence together, Professor Neugebauer concludes that kekros refers to 1/60.

One particularly impressive use of base 60 appears in an Amharic manuscript. The manuscript gave the mean value of a synodic month as 29:31,50,7,57,30d. How should this be interpreted? This shows that the value was $29 + 31/60 + 50/60^2 + 7/60^3 + 57/60^4 + 30/60^5$ days in length. This is clearly a complex calculation.

CHAPTER 8: SHARED GAMES AND MATHEMATICAL CONCEPTS IN ETHIOPIA AND THE REST OF AFRICA

Mancala from Ethiopia to the rest of Africa

The oldest known evidence of the Ancient African board game, Gebet'a or "Mancala" as it is more popularly known, comes from the ancient site of Yeha in Ethiopia. The recovered artefact may well date back to around 700 BC. The game challenges players to strategically capture a greater number of stones or pebbles than one's opponent. The game usually consists of a wooden board with 2 rows of 6 holes each, and 2 larger holes at either end. In some versions of the game, players are allowed to make backward moves. Professor Zaslavsky likens backward moves to negative numbers and forward moves to positive numbers.

More advanced versions of this game are found in Central and East Africa, such as Bao, Igisoro and Omweso. It was part of the accession ritual in the old Ugandan Kingdom of Buganda that a newly enthroned king must play Omweso against the Prime Minister to symbolise that a King must be able to outwit and defeat his subjects by strategy. They usually have 4 rows of 8 holes each. Each of the two players has 32 beans to play with. Lela, as the game was called in central Africa, was a game supposedly popularised by King Shamba Bolongongo of the Kuba Kingdom of Congo, around 1610. Dissuading his citizens from playing gambling games, he proposed Lela as an intellectual alternative. In the Ashanti Empire, royalty played a similar game was called Wari or Oware on golden boards made in the shape of a royal stool! In the Songhai Empire it was called Sudanese Chess.

The game was even brought over to Surinam, South America, by enslaved Africans where it is called Adji Boto.

Professor Zaslasky summed up the pedagogical importance of the game when she wrote:

> It is incredible that these African games were actually discouraged by the colonial education authorities in favour of ludo, snakes-and-ladders, and similar games of European origin. I agree wholeheartedly with John B. Haggarty who writes: 'Kalah [an American version, commercially produced] is the best all round teaching aid in the country ... In addition to its value ... as a

means of developing the intuitive abilities so useful to problem-solving, there is another outcome equally valuable. This outcome is the recognition of the close identification of the game throughout the history of civilization with the development of systems of numeration and the concept and ideas of number.'

Setati and Bangura cite the research of Johnson Ihyeh Agbinya who points out that these African board games involve shift, add and subtract operations. Moreover many of the moves can be represented with a function in which the numerical values increase by one at a time except one value that is decreased to zero entirely.

Fractal Geometry in Africa

A fractal is a geometric shape. They key feature is that it can be divided into pieces in a way that each piece is a reduced copy of the whole at ever diminishing scales. The term fractal is new but fractal shapes have been used in African cultures for centuries. This is particularly true of buildings, art, textiles and other forms of decoration. Moreover fractal shapes exist in nature and have an "organic" beauty. Delightful images of fractals were at one time very popular as screensavers on computers.

Professor Ron Eglash is the leading authority on African fractals and documented fractal patterns in cornrow/canerow hairstyles, weavings, and the architecture of villages and palaces, as well as many forms of African art, craft and even some forms of music. One the importance of fractals, Professors Setati and Bangura write:

> What Eglash teaches in the 14 chapters is that elaborate cornrow braids on an African woman's head, for example, can be viewed as more than an affinity with culture or a fashion statement. The intricate patterns are also useful for learning about African fractals--geometric patterns that are repeated on smaller and smaller scales to produce intricate designs that are beyond the scope of classical or Euclidean Geometry. Fractal Geometry has emerged as one of the most exciting frontiers in the fusion between Mathematics and Information Technology. Fractals can be observed in many of the swirling patterns produced by computer graphics, and they have become a vital tool for model[l]ing in the natural sciences. While Fractal Geometry can allow one to get into the far reaches of high tech science, its patterns are surprisingly common in traditional African designs. Also, some of the basic concepts in Fractal Geometry are fundamental to African knowledge systems: quantitative techniques, symbolic systems, engineering, architecture, games, traditional hairstyling, textiles, sculpture, painting, carving, metalwork, and religion.

Why is this of key importance? Professor Eglash explained its importance in a famous TED lecture:

The most complex example of an algorithmic approach to fractals that I found was actually not in geometry, it was in a symbolic code, and this was Bamana [also called Bambara] sand divination. And the same divination system is found all over Africa. You can find it on the East Coast as well as the West Coast, and often the symbols are very well preserved, so each of these symbols has four bits--it's a four-bit binary word -- you draw these lines in the sand randomly, and then you count off, and if it's an odd number, you put down one stroke, and if it's an even number, you put down two strokes. And they did this very rapidly, and I couldn't understand where they were getting--they only did the randomness four times --I couldn't understand where they were getting the other 12 symbols. And they wouldn't tell me ... And one of them said, 'Come here. I think I can help you out here' ... And finally, he revealed the truth of the matter.

Eglash continues:

And it turns out it's a pseudo-random number generator using deterministic chaos. When you have a four-bit symbol, you then put it together with another one sideways. So even plus odd gives you odd. Odd plus even gives you odd. Even plus even gives you even. Odd plus odd gives you even. It's addition modulo 2, just like in the parity bit check on your computer. And then you take this symbol, and you put it back in so it's a self-generating diversity of symbols. They're truly using a kind of deterministic chaos in doing this. Now, because it's a binary code, you can actually implement this in hardware--what a fantastic teaching tool that should be in African engineering schools.

This was not the only powerful idea. Eglash had more to say:

And the most interesting thing I found out about it was historical. In the 12th century, Hugo of Santalla brought it from Islamic mystics into Spain. And there it entered into the alchemy community as geomancy: divination through the earth. This is a geomantic chart drawn for King Richard II in 1390. Leibniz, the German mathematician, talked about geomancy in his dissertation called *De Combinatoria.* And he said, 'Well, instead of using one stroke and two strokes, let's use a one and a zero, and we can count by powers of two.' Right? Ones and zeros, the binary code. George Boole took Leibniz's binary code and created Boolean algebra, and John von Neumann took Boolean algebra and created the digital computer. So all these little PDAs and laptops--every digital circuit in the world--started in Africa.

How were fractals used in Ethiopia?

Ethiopian church crosses and even the famous churches in the mediaeval Ethiopian city of Lalibela use fractal geometry, through and through. Some designs used circles, squares, rectangles over and over again at ever diminishing scales.

Figure 25. Graphics showing how Ethiopian church cross designs were built using fractals (after Ron Eglash).

CHAPTER 9: MATHEMATICS IN WEST AND CENTRAL AFRICA

Historical Introduction

A number of writers have found the non-Islamic civilisations of West and Central Africa to be interesting sources of practical mathematical knowledge. Very little of it is pure mathematics and most of it, until the 1970s, was not even considered mathematics at all. The Yoruba Kingdoms of what is now south western Nigeria flourished in mediaeval times but their principal city of Ife was founded in the sixth century AD. Another great state existed nearby called the Benin Kingdom. Its first monarchs began to rule around 900 AD. Evidence from the Ghana region comes from the Begho culture from the fourteenth century and the Ashanti Empire of the eighteenth and nineteenth centuries. Evidence also comes from Central Africa such as the Kuba Kingdom of Congo, the Shongo Kingdom nearby and the Tchokwe culture of Angola.

Introduction to indigenous West and Central African Numerical Systems

According to Professor Théophile Obenga, a number of African peoples had sophisticated numerical systems of their own. For example:

o The Yoruba of Nigeria have an indigenous word for 1,000,000 i.e. *egbeeberun*.

o The Ganda (or Buganda) of the Great Lakes region have an indigenous word for 20,000,000 i.e. *ebutabalika*.

o The Bakongo of Kasai/Sankuru have an indigenous word for 1,000,000 i.e. *losenene*.

o The Duala of Cameroon have an indigenous pair of words for 3,000,000 i.e. *lodun lolálo*.

o The Fang of Equatorial Guinea have an indigenous phrase for 20,000,000 i.e. *bidudum mewom mebe*.

Professor Obenga regards these examples as evidence that suggests (i) these societies worked with very high numbers, (ii) they developed suitable indigenous terminologies for them, and (iii) they could thus conceptualise numerically. It must be pointed out that is likely that neighbouring cultures borrowed intellectual terms from each other. Moreover, the presence of Islam in many parts of Africa meant that people also borrowed terms from Arabic. This meant that some African societies would have had access to a wide variety of mathematical ideas, some wholly indigenous, others borrowed from neighbours, and others from the Islamic world, of which Africa was an integral part.

According to the author of the book *After God is Dibia,* the Igbo of what is today Eastern Nigeria, have long had indigenous linguistic terms enabling them to count to a billion. According to Professor Zaslavsky, the Igbo had a special term for 20^2 or 400 which is *nu;* 400^2 or 160,000, is expressed as *nnu khuru nnu,* translated as "400 meets 400." They also had a term for infinity, *pughu,* which means an "uncountably" large number.

The Bram and Mankanye people of Guinea-Bissau have interesting terms for some of their numbers. Five is denoted by the same word for "hand," ten is "two hands," nine is "hand and hand less one," and 19 is expressed as "two hands and hand and hand less one." However 12 is expressed as "six times two." They carry the idea further with 24 = 6 x 4. This means to say "24" in these languages, one must say the equivalent of six multiplied by four! Their neighbours, the Balante, employ six to delineate the numerals from seven to 12. Thus 7 = 6 + 1, 8 = 6 + 2, etc. For the Ga people of Ghana, both seven and eight are based upon six: 7 = 6 + 1 and 8 = 6 + 2.

The Huku of Central Africa based their numerical system on numbers four and six. Thus:

7 = 6 + 1
8 = 6 + 2
9 = (2 x 4) + 1
10 is an independent word
13 = 12 + 1
16 = (2 x 4) x 2
17 = (2 x 4) x 2 + 1
20 = 2 x 10

Yoruba Numerals

If you thought having to calculate [(2 x 4) x 2 + 1] before saying 'seventeen' was complicated, imagine having to calculate [(200 x 3) - (20 x 4) + 5] before being able to say 'five hundred and twenty five.'

According to mathematician, Professor Claudia Zaslavsky, author of the classic *Africa Counts,* you 'must be a mathematician' to use the Yoruba number system. To illustrate this 170 is the Yoruba equivalent of (20 x 9) - 10, 185 = (200 - 10) - 5, and 5000 = 20 x 25. For this reason, learning Yoruba numerals have a pedagogical value in Nigeria and among some of the African Americans. Educators see the value in teaching this system to pupils since it gets them to use arithmetic in just being able to express the numbers. This system has been in use for hundreds of years and may well date back to the glory days of the Kingdom of Ife which was founded perhaps as early as the sixth century.

The Yorubas evolved a complicated numerical system that often involves subtraction and multiplication to express a single number. The Yoruba phrase for three hundred and fifteen is *orin* (which means 20 x 4) *din nirinwo* (from 400) *odin marun* (less 5), which, in mathematical symbols, becomes 400 - (20 x 4) - 5 = 315. The English equivalent 'three hundred and fifteen' is simply (3 x 100) + 15 = 315, making use of multiplication and addition, but no use of subtraction. Many centuries ago, however, when Roman numerals were used across Europe, subtraction was also used. The Roman IV, for example, is 5 - 1, and IX is 10 - 1.

The Yorubas had separate terms for one to ten: i.e. *ookan, eeji, eeta, eerin, aarun, eefa, eeje, eejo, eesan* and *eewaa.* From ten to fourteen, the Yorubas use addition. Their phrase for eleven, for example is *ookan laa,* which is 10 + 1 = 11. The other numbers mean 10 + 2, 10 + 3 and 10 + 4.

However for fifteen, they say *eedogun,* which derives from *arun* (five) *din ogun* (from twenty) or 20 - 5 = 15. The other numbers from sixteen to nineteen are 20 - 4, 20 - 3, 20 - 2, and 20 - 1. Between twenty and thirty, there is a similar pattern--addition is used from twenty-one to twenty-four and subtraction is used from twenty-five to twenty-nine. For example, twenty-two becomes 20 + 2 and twenty six becomes 30 - 4.

The Yorubas count in base 20, in contrast to what we use today and other systems in Africa--base 10. Consequently twenty, and numbers that are multiples of twenty, are important in their system. The word for twenty is *ogun.* The word for forty is *ogoji,* which is derived from *ogun* (twenty) and *eeji* (two). Sixty is *ogota,* derived from *ogun* and *eeta* (three), and eighty (*ogorin*) comes from *ogun* and *eerin* (four). In mathematical symbols 40 = 20 x 2, 60 = 20 x 3 and 80 = 20 x 4. They have special names for important base 20 numbers, such as *igba* (200) and *irinwo* (400), just as the English have special names for important base 10 numbers such as 100 (hundred).

For numbers fifty, seventy and ninety, subtraction is used. The term for fifty is *aadota.* It comes from *ogota,* which, as we have seen, means 20 x

3, and *laa* which is ten, but in this context means minus ten. Fifty is therefore (20 x 3) - 10, seventy is (20 x 4) - 10 and ninety is (20 x 5) - 10.

The Yoruba terms for numbers forty-five to forty-nine are complicated, as are sixty-five to sixty-nine and eighty-five to eighty-nine. For example, 46 = (20 x 3) - 10 - 4, 67 = (20 x 4) - 10 - 3 and 88 = (20 x 5) - 10 - 2.

The Yoruba number system had terms for unit fractions 1/2, 1/3, 1/4 and 1/5, etcetera. The Yoruba divination system called the *Odu Ifa* had phrases for 4^2 (meaning 4 x 4 = 16), 4^3 (or 4 x 4 x 4 = 64), and 4^4 (or 4 x 4 x 4 x 4 = 256). The Yorubas also had an indigenous notion of infinity and, as we have seen, they had a word for million (*egbeeberun*).

Geometry in West and Central Africa

It may seem the height of disrespect to compare the old and exquisite Benin Bronzes or the fabulous Congolese Textiles to wallpaper, but there is a connection. Wallpaper, despite its apparent blandness, is actually a treasure trove of applied mathematical techniques. Typical wallpaper designs are based on a single idea or motif that is repeated across the paper using two geometric operations, translation and reflection.

Translation is where a motif is repeated by moving it in a straight line from its original position to position one, two, three, four, etcetera. Reflection is where a motif is repeated as if it were a mirror image of the original. Reflection can take place along a vertical axis, making the left

Figure 26. Superb embroidered raffia cloth from the Kingdom of Kuba (now modern Congo). Eighteenth century. The interlacing *mbolo* pattern is similar to those traditionally drawn by Kuba children as a childhood game.

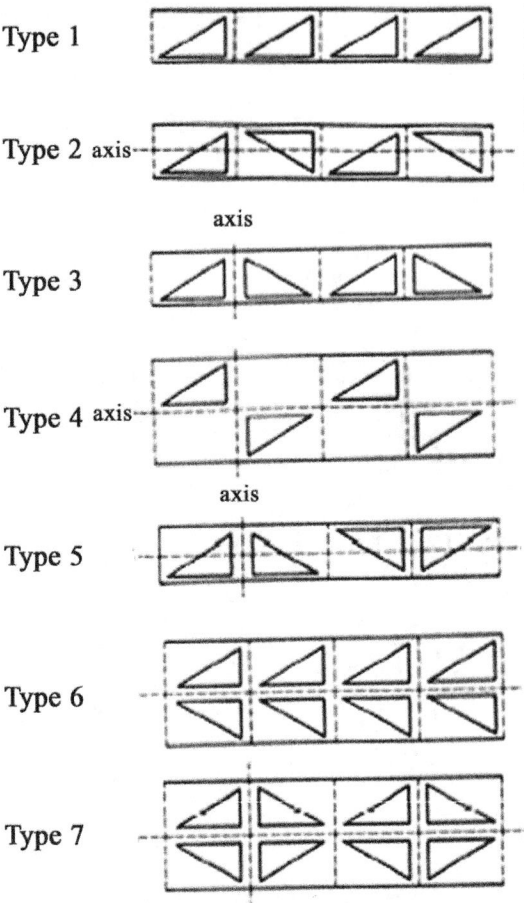

Type 1

Type 2 axis

axis

Type 3

Type 4 axis

axis

Type 5

Type 6

Type 7

Figure 27. Professor Claudia Zaslavsky used this schema to show the different geometric patterns that mathematicians use to classify geometric designs that repeat themselves in one direction.

hand side of the original motif, the right hand side of the reflected motif. It can also take place along a horizontal axis, making the bottom of the original motif the top of the reflected motif.

There are 24 different combinations of translation and/or reflection (including rotating the motif by 60 degrees, 90 degrees, 120 degrees, 180 degrees or 360 degrees) that can be used to cover a wall. The proof that there were only 24 such techniques allegedly came from a Russian scholar called Evgraf Federov in 1891. An analysis of Islamic art, however, proved that the Islamic world had long made use of all 24 techniques. This raised the question of whether this knowledge existed in West and Central African arts and crafts.

Dr Donald W. Crowe and Professor Zaslavsky made independent studies of the Benin Bronzes of the Nigeria region and the Bakuba Textiles from

Figure 28. The Benin Bronze masterpieces of the sixteenth century show all seven geometric patterns that repeat in one direction. Dr Donald W. Crowe has selected designs from Benin art to illustrate each example.

the Congo region. Dr Crowe also made a study of the Begho smoking pipes of the Ghana region. Some of the Begho evidence dates from the fourteenth century onwards. The Benin masterpieces were mostly from the sixteenth century. The Congolese crafts were mostly from the eighteenth century. Dr Crowe concluded that 14 of the 24 possible mathematical techniques were used in the Begho pipes, 17 of the 24 possible mathematical techniques appeared in the Benin Bronzes, and 'at least' 19 of the 24 techniques appeared in the Bakuba Textiles.

There are two types of patterns, strip and plane patterns. Strip patterns, also called frieze patterns have only one direction of repeat. Plane patterns, also called wallpaper groups, have two directions of repeat. In each case Dr Crowe found all of the 7 possible strip patterns. However, he found only 7 of the possible 17 plane patterns in the Begho Pipes, 10 of the possible 17

Figure 29. A selection of delicate Bakuba textiles chosen by Professor Zaslavsky to illustrate a variety of geometric patterns.

plane patterns in the Benin Bronzes, and 12 of the possible 17 plane patterns in the Bakuba Textiles.

The two-colour plane patterns in Bushongo designs existed in the structure of woven mats, the decorative pattern on the border of loincloths, and as patterns on women's bark cloth. Of the polygonals used, were light and dark rectangles, squares, triangles and rhombi sewn together.

The Ashanti Empire of what is today Ghana and a portion of the Ivory Coast, brought pottery to a high level of perfection. Some of the artefacts are richly decorated with geometric designs. Some jars are decorated with an 8-fold symmetry design and a strip pattern. Other Ashanti artefacts interwoven with geometrical ideas and experimentation are plaster (wall decoration), smith work (e.g. blades of swords), exquisite jewellery, and the weaving and decoration of cloth. To weigh gold dust, the Ashanti traditionally used weights often with geometric or animal shaped forms. The best known West African fabric is the multi-coloured *kente* cloth that was woven by the Ashanti in Ghana and also the Ewe in Togo. *Kente* weavers use horizontal looms to produce long and narrow strips of cloth that were sewn together to form square or rectangular pieces used to make costly and majestic robes. Some of these clearly display two-colour patterns that demonstrate translations, reflections or rotations which reverse the colours of a particular pattern.

Polygonal shapes are even important in making a drum. To get a good tone from the instrument, one must ensure that the covering is fixed evenly to the drum's wall. Thus, the pins must be equally spaced. Stated mathematically, the holes in which the pins are nailed must constitute a regular polygon. The details and complexity of the polygon will vary with the size and pitch of the drum.

Figure 30. Splendid design from an Ashanti jar showing 8-fold symmetry (after Gerdes).

Other uses of geometry are demonstrated by a game traditionally played by Shongo children in the Congo region. According to mathematician Professor Beatrice Lumpkin, the algorithm involved drawing complex networks with a continuous line or path, without taking one's pen or pencil off the page and without tracing the same line twice. A European mathematician, Leonhard Euler (a contemporary of Christian Goldbach mentioned in the *Introduction*) developed solutions to this problem in 1735. He also established a new branch of topology called network analysis. However, according to Professor Lumpkin, it is likely that the Shongo games are much older than the time of Euler.

The Tchokwe of Angola have a tradition of drawing designs in the sand that were traditionally drawn alongside storytelling. As with the Bushongo networks, the objective was to trace designs without breaking or retracing the same line. Moreover, this tradition dated back centuries. A visitor to the Angola region in the seventeenth century, Cavazzi, sketched a Tchokwe scarf that had a network design on it.

Professor Paulus Gerdes, author of the superb *Geometry From Africa: Mathematical and Educational Explorations* and *Drawings from Angola: Living Mathematics,* shows that the Tchokwe designs show rotational symmetries of 90 and 180 degrees and often have one or more axes of symmetry.

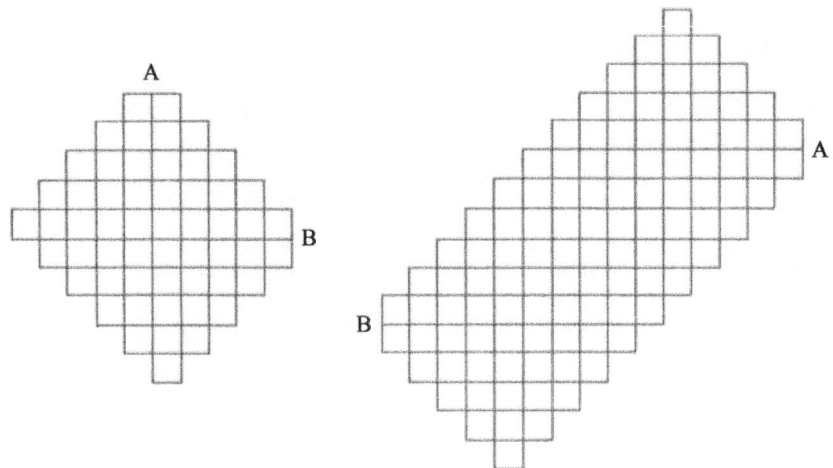

Figure 31. Typical networks traditionally drawn by Shongo children as a game. In order to draw these networks, one must start at the entry points A or B.

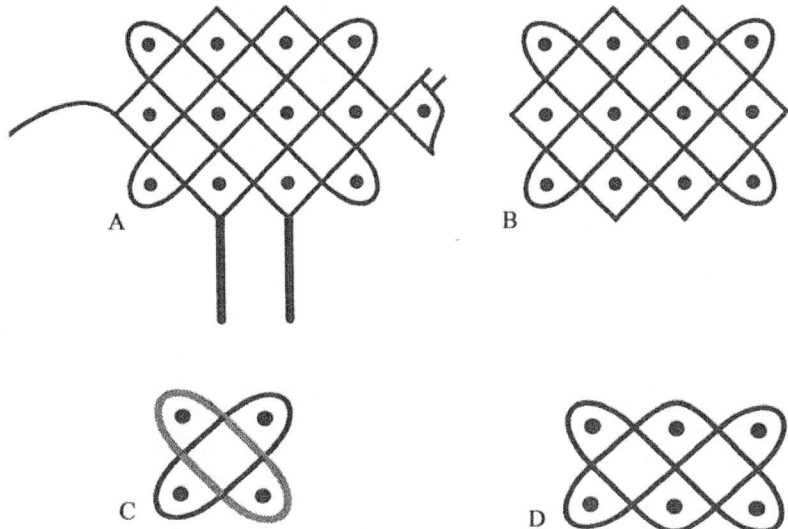

Figure 32. The Tchokwe antelope design is shown here with the head, tail and legs (see A) or without (B). C is a 2 by 2 design requiring two complete lines to draw. This shows that the greatest common divisor of 2 and 2 is 2. D is a 2 by 3 design requiring one complete line to draw. This indicates that the greatest common divisor of 2 and 3 is 1.

He shows that the Tchokwe designs can be classified by the tracing algorithm. Gerdes distinguishes six different sets of geometric algorithms traditionally used by the Tchokwe to trace their designs. One of these six sets was the plaited-mat design. Gerdes illustrates this set with many examples of Tchokwe designs but the key ones were the traditional sand drawings of antelopes and lionesses.

Ignoring the head and the legs, the antelope was traced around 12 points consisting of 3 rows of 4 points. The lioness was traced around 30 points consisting of 3 rows of 10 points.

Gerdes shows that to trace designs like this, there is a clear relationship between the number of points and how many complete lines one would need to trace around all of the points. He shows that if the design was to be traced around 4 points consisting of 2 rows of 2 points, one would need TWO complete lines to trace the design. If the design was to be traced around 10 points of 2 rows of 5 points, one would need ONE complete line to trace the design. Also, if the design was to be traced around 18 points of 3 rows of 6 points, one would need THREE complete lines to trace the design. Finally, if the design was to be traced around 20 points of 4 rows of 5 points, one would need ONE complete line to trace the design.

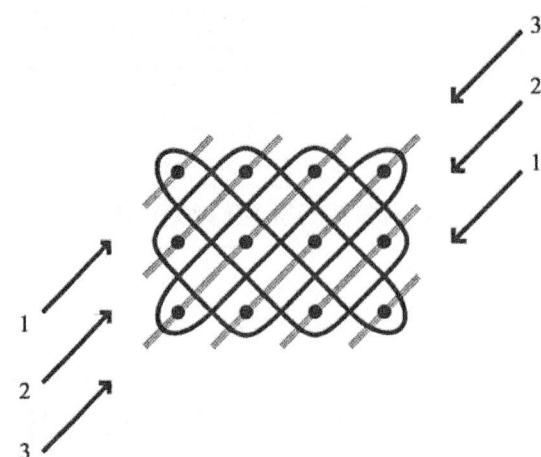

This is significant because the number of complete lines to trace the antelope or the lioness designs represents the greatest common divisor of the number of rows and number of columns of points. Thus the greatest common divisor of 2 and 2 is 2. The greatest common divisor of 2 and 5 is 1. The greatest common divisor of 3 and 6 is 3. The greatest common divisor of 4 and 5 is 1. Thus the traditional tracing algorithm for the antelope or the lioness used by the Tchokwe encompasses the greatest common divisor.

Gerdes demonstrates that the antelope and lioness designs show other mathematical data. The antelope design has 3 rows of 4 points and indicates that 3 x 4 = 12. The points can also be counted along the diagonals from left to right as 1, 2, 3, 3, 2 and 1. This indicates that 3 x 4 = 12 = (1 + 2 + 3) + (3 + 2 + 1). The second side of the equation can be rearranged as (1 + 2 + 3) + (1 + 2 + 3) = 2 x (1 + 2 + 3).

Moreover:
2 x (1 + 2 + 3) = 3 x 4 = 12

Therefore:
1 + 2 + 3 = (3 x 4)/2 = 6

Gerdes concludes that this and other antelope and lioness designs suggest that: n x $(n + 1)$ = 2 x $(1 + 2 + 3 + ... + n)$ also n x $(n + 1)/2$ = $1 + 2 + 3 + ... + n$.

In the antelope example above n is 3. Thus 3 x (3 + 1) = 2 x (1 + 2 + 3) also (3 x 4)/2 = 1 + 2 + 3.

Why is this useful? Suppose a mathematics problem required you to add all numbers from 1 to 3. The answer is 6. We could add up $1 + 2 + 3$ or we could use the formula where $n \times (n + 1)/2$ or $(3 \times 4)/2 = 6$. The advantage of using this formula becomes readily apparent if the mathematics problem required you to add all numbers from 1 to 100. We could laboriously add up $1 + 2 + 3 + ... + 99 + 100$ or we could use the short cut $n \times (n + 1)/2$ or $(100 \times 101)/2 = 5050$.

CHAPTER 10: MATHEMATICS IN THE WEST AFRICAN SUPERSTATES

Historical Introduction

The desert belt of West Africa was home to three great civilisations: Ancient Ghana (not to be confused with modern Ghana), Mediaeval Mali and the Songhai Empire. Flourishing as imperial systems these polities held sway of vast swathes of West Africa from around 700 to 1591. Most writers treat these three cultures as one topic and I shall do the same. Centres of Islamic learning were established in the West African desert cities of Kumbi-Saleh, Walata, Timbuktu, Djenné, Gao and Katsina. Some one million old manuscripts have survived. Highly advanced non-Islamic learning also flourished in the initiation societies of the Dogon and the Bambara.

Geometry

The earliest surviving evidence of mathematical endeavour in the West African superstates comes from the eleventh up to the fifteenth centuries by a people now known as the Tellem. Archaeologists and textile experts who have examined the surviving Tellem textiles assert that no other region in the world has had such a great variety of linear and geometrical patterns in cotton fabrics by means of a single colour. They used the only dye then available to them: i.e. indigo.

Professor Paulus Gerdes, a learned authority on African geometry, has studied and analysed patterns found on preserved fragments of tunics, sleeves, coifs and caps, woven in plain weave. This is where the horizontal and vertical threads cross each other one over, one under. The weavers alternated groups of natural white cotton threads with groups of blue, indigo-dyed, threads.

Gerdes saw patterns where from left to right, six vertical white threads were followed by four blue threads; from top to bottom, and three horizontal white threads were followed by three blue threads. These yield a plane pattern giving a basic rectangle with dimensions ten (= 6 + 4) by six (= 3 + 3), or (6 + 4) x (3 + 3). Gerdes states that generally, the dimensions of the shapes that emerge from the weave patterns can be expressed as $(m + n)$ x $(p + q)$, where $m, n, p,$ and q are natural numbers.

The Tellem weavers experimented with dimensions and found relationships between the dimensions and the symmetry properties of the patterns that resulted. In particular, the variation among the plain weave fragments suggests that the weavers were familiar with the effect on the patterns of the selection of even and odd dimensions, in addition to how these dimensions $(m + n)$ and $(p + q)$ were produced.

The Tellem patterns from the eleventh and twelfth centuries feature woven rectangles followed by fragments of respective plane patterns, which are two-colour patterns in the sense that for each there is a rigid motion of the plane--translation, rotation, reflection--that reverses the blue and white colours.

Other early evidence of geometric thinking comes from linguistic terms that have survived in the Soninke language, the imperial language of Ancient Ghana, the first of the powerful West African superstates. The terms *koore* and *gogomme* means circle, circumference and circular area. *Fana* means circular shape, circular object or a disk. *Kutura* means sphere. *Guniye* means a curve, bow shape or an arc of circumference. Finally the term *morogomoroye* means cylinder. Moreover, these terms were NOT derived from practical and concrete realities such as the words for calabash, bowl, egg or earthenware pots. These terms therefore reflect theoretical speculation and not just concrete thinking.

Magic Squares

How is it possible to arrange numbers into a table using each number only once so that each row, each column, and the two diagonals all add up to the same number? This question had intrigued Chinese scholars for thousands of years. However, West African scholars were also interested in this puzzle. Not only did they produce solutions to this problem, they even wrote examples as good luck charms and benedictions!

Archaeologists working in the Malian city of Djenné revealed some interesting information reported by Karen E. Lange:

> In the base of a wall from about A.D. 1400 they found fragments of a type of bowl the Djennenké [i.e. people of Djenné] still place in foundations for protection. One fragment carried magical grids or squares; another was inscribed with a benediction in Arabic; the third had the date 512--or, adjusting from the Islamic calendar, A.D. 1118.

Figure 34. Page from Muhammad Ibn Muhammad's *A Treatise on the magical use of the letters of the alphabet*, 1732.

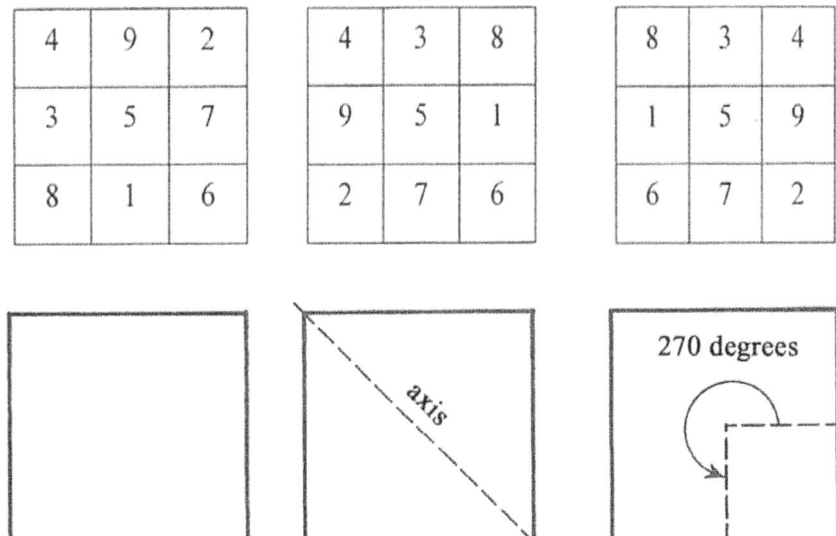

Figure 35. This is a 3 x 3 magic square described by Ibn Muhammad. The magic constant is 15. He manupulates these squares by reflecting them and rotating them.

Thus, the people of Djenné were familiar with magic squares at least as early as 1400 or even as early as 1118 AD. Scholars in Borno, a neighbouring state to the east of Mali and Songhai, were teaching on magic squares as late as the eighteenth century.

What are magic squares? A magic square is a mathematical recreation or game. It is constructed by arranging numbers into a table where each row, each column, and the two diagonals, must add up to the same number called the magic constant. A Hausa scholar from the University of Katsina, Ibn Muhammad, published a book in 1732 with examples in it. Professor Claudia Zaslavsky made a special study of this manuscript, originally called: *A Treatise on the magical use of the letters of the alphabet.*

Ibn Muhammad worked with three order squares (i.e. 3 x 3 = 9 squares in total), five order squares (i.e. 5 x 5 = 25 squares in total) right up to eleven order squares (i.e. 11 x 11 = 121 squares in total).

He also demonstrated how a given magic square can be reflected about the vertical axis, the horizontal axis, and about the two diagonals. Moreover, he showed how a given square can be rotated through 90 degrees, 180 degrees and 270 degrees.

13	25	7	19	1
17	4	11	23	10
21	8	20	2	14
5	12	24	6	18
9	16	3	15	22

Figure 36. Ibn Muhammad constructed this 5 order square using the knight's move (or -1, -2) from numbers 1, to 2, to 3, etcetera.

How did Ibn Muhammad solve the problem? One solution was similar to using the knight's move in chess. Starting from the square in the top right hand corner, he placed a '1'. Moving one square to the left and two squares down (i.e. -1, -2), he placed a '2'. Moving one square to the left and two squares down (-1, -2), he placed a '3'. However, to make the same (-1, -2) move would move us off the square. Consequently note where Ibn Muhammad placed the '4'. The '4' is also an example of one square to the left and two down, but imagine that the square has been wrapped around a cylinder so that the top of the square is in contact with the bottom of the same square! Once this is understood, the placing of the '4' makes sense. It is still following the (-1, -2) rule.

Ibn Muhammad demonstrated a second method to construct a 5 order square. First he wrote the numbers 1 to 25 from right to left. Secondly, he superimposed a second square rotated by 45^0 on the same diagram. Thirdly, he used the information from this second square to begin constructing a magic square. Fourthly, he completed the magic square.

Professor Claudia Zaslavsky, mentioned earlier, shows that a magic square of odd number n (3, 5, 7, etc.), consists of a square array of numbers from 1 to n^2, then the magic constant will be equal to $n(n^2 + 1)/2$. For example, if we try to construct a seven order magic square ($7 \times 7 = 49$), then $n = 7$. This means that the numbers used will be 1 to 7^2 (= 49) and the magic constant will be $7(7^2 + 1)/2 = 175$.

Commenting on this, Professor Finch, an authority on African science history, noted that this shows 'the "algebraic" quality of magic squares and why a sound knowledge of number theory is important in their creation.' This raises the question: Did West Africans only get as far as algebra?

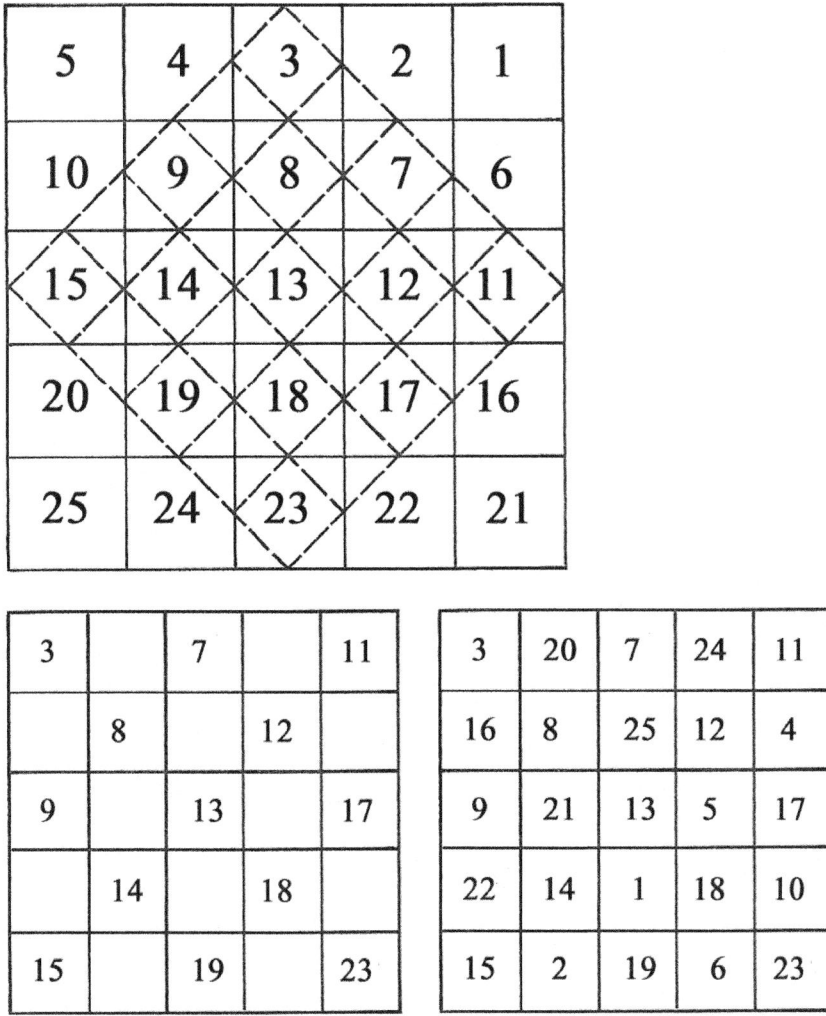

Figure 37. Ibn Muhammad constructed this 5 order square in stages. *Top:* He wrote the numbers 1 to 25 from right to left and superimposed a second square at a 45⁰ angle on the same diagram. *Bottom left:* He used the information from this 45⁰ square to begin constructing a magic square. *Bottom right:* He completed the magic square.

A *New Scientist* article had plenty to say on the practise of mathematics at Timbuktu which answers this question:

> [T]he Muslim scholars of Timbuktu would ... have had particular reasons to be interested in astronomy. First is the requirement for Muslims to pray, and to

orient their mosques in the direction of Mecca. To achieve this, they needed to develop algorithms and instruments to determine the exact position of both Mecca and Timbuktu. There was also the need to determine exact times for prayers at sunrise, noon, afternoon, sunset and evening. The scholars found ancient Greek methods of doing this very cumbersome, the researchers say, but Muslim astronomers devised easier solutions by inventing the cosine, tangent, cotangent, secant and cosecant functions of trigonometry.

Moreover, mathematics was one of the liberal arts taught at the Songhai universities. The scholars, Hunwick and Boye, state that Timbuktu scholars purchased and copied manuscripts on geometry and calculus. Finally, Professor Henry Louis Gates, the famous African American academic, drew attention to a surviving manuscript in a Timbuktu library on mathematical accounting.

West Africans (like Africans elsewhere) made practical use of fractal geometry. The Rao Pectoral was a magnificent golden artefact buried with a vassal prince of Ancient Ghana from the twelfth or thirteenth centuries AD. The pectoral uses the circle over and over at different scales, offering a pleasing aesthetic.

Finally, the scholars of Hausaland (northern Nigeria) and Borno were studying arithmetic from the eleventh and the fifteenth centuries respectively. They used arithmetic to calculate inheritance, collecting and

10	45	44	7	11	12	46
9	19	34	17	20	35	41
8	18	24	23	28	32	42
49	37	29	25	21	13	1
48	36	22	27	26	14	2
47	15	16	33	30	31	3
4	5	6	43	39	38	40

Figure 38. A seven order magic square on a Hausa amulet was described in an 1896 book published in Cambridge. It is a bordered magic square where even if one successively removes border after border, the square remains magic! The magic constant is 175. Once a layer is removed it becomes 125. Once another layer is removed it becomes 75. Once this layer is removed, the final number left is 25.

distributing *zakat* (i.e. charity), in business, and in land surveying. They are known to have consulted Coptic Solar Calendars in determining their economic activities. An old Arabic manuscript written in a Sudanic (i.e. West African) script, but probably composed in mediaeval Egypt turned up in the ancient Nigerian city of Bauchi in 1973. The book contained mathematical charts dealing with agronomic activities such as the right time of harvest; the various directions of the wind; time of germination; and the seasons during which insects appear. A conversion table to lunar months was also made at the beginning of the book as a guide for the users of the chart. There were also geometric texts. The great nineteenth century ruler Sultan Muhammad Bello had a family copy of Euclid's *Elements* in Arabic.

CHAPTER 11: MATHEMATICAL IDEAS IN TRADITIONAL AFRICAN KNOW-HOW, CRAFTS AND DESIGNS

Geometry in Mozambique

Professor Paulus Gerdes, a leading authority on African mathematics, reveals that traditional African know-how often demonstrates sophisticated geometrical knowledge. He demonstrates this by using examples of African wall decorations, rolled up mats, woven knots, woven pyramids, square mats, plaited mats, and plaited strips. However, two particularly impressive sets of examples come from traditional house building techniques and hexagonal weaving techniques.

The peasantry in Mozambique traditionally used either of two methods to lay out a house on a rectangular plan with the corners being right angles.

The first method made use of 4 long bamboo sticks. Sticks 1 and 2 were of equal length. Sticks 3 and 4 were of equal length to EACH OTHER, but SHORTER than 1 and 2. The sticks were arranged on the ground into a quadrilateral where 1 and 2 were opposite each other and 3 and 4 were opposite each other. These 4 sticks ultimately became the boundary of the house plan. To make the four corners right angles, the peasants traditionally used a rope to measure the diagonals. They would adjust the positions of the 4 bamboo sticks until the 2 diagonals were of equal length.

The second method made use of 1 bamboo stick and 2 pieces of rope of equal length. The 2 pieces of rope were tied to each other at their exact mid points. They were also connected to each end of the bamboo stick. The stick was laid on the ground and became one of the sides of the house. The two ends of the stick became two vertices of the house plan. Each piece of rope was pulled

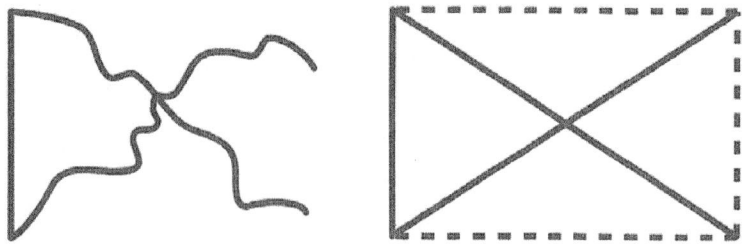

Figure 39. The left hand diagram shows the two pieces of rope attached to a bamboo stick. The right hand diagram shows the rope being pulled tightly (after Gerdes).

tightly so that each one became a straight line. The ends of each of the two pieces of rope became the other two vertices of the house plan.

These practical house plan methods demonstrate know-how equivalent to Euclidean theorems that state that parallelograms with congruent diagonals are rectangles. Moreover, quadrilaterals with congruent diagonals that intersect at their midpoints are rectangles.

Gerdes reveals that this know-how can be used to formulate alternative rectangle axioms such as: "If a quadrilateral ABCD, AD = BC, AB = DC and AC = BD, then LA, LB, LC and LD are right angles." Also: "If a quadrilateral ABCD, M = [the intersection between] AC [and] BD and AM = BM = CM = DM, then LA, LB, LC and LD are right angles, AD = BC, and AB = DC."

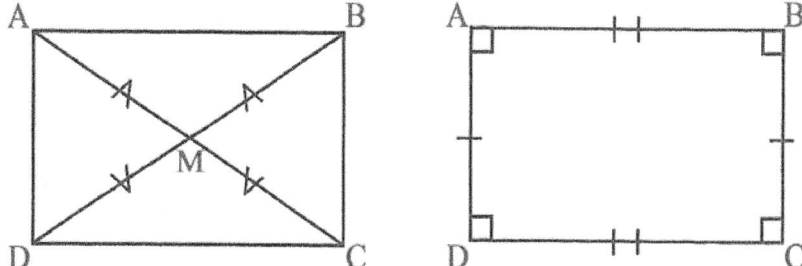

Figure 40. If M is the intersection between AC and BD and AM = BM = CM = DM, then LA, LB, LC and LD are right angles, also AD = BC, and AB = DC (after Gerdes).

In Mozambique, artisans traditionally used hexagonal weaving techniques to make hats, handbags, transportation baskets and fish traps. If the strips used for the craft making were of equal width, they were traditionally plaited at 60 degrees to each other. Other angles would have been used if the strips were of different widths. The 60 degree angles meant that the finished craft pieces contained equilateral triangles, rhombi, trapeziums, and regular hexagons. However, the artisans found that by including pentagons into the design that replaced some of the hexagons, it was possible to create curvature in the craft pieces, giving a sense of three dimensions.

Professor Gerdes goes further and suggests that the degree of curvature could be increased by including squares and triangles replacing the hexagons. Moreover, he shows that these craft techniques could be used to make or model polyhedra. This raises the question: Did the traditional artisans of Mozambique know of the five regular polyhedra (also called the Platonic solids)? In my opinion, the answer at the present state of knowledge is 'no.' However, it is clear that the artisans had knowledge that represents some of the steps that would have led to the discovery of the Platonic solids.

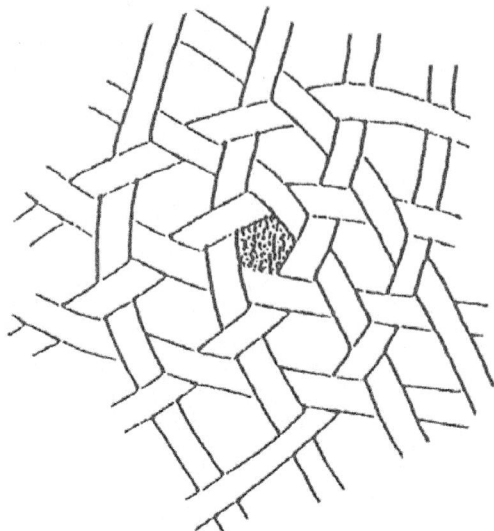

Figure 41. Including pentagons into hexagonal weaving designs, replacing some of the hexagons, creates curvature in the craft pieces and leads to three dimensionality (after Gerdes).

Toothed Squares

Toothed square designs exist in many parts of Africa. Professor Paulus Gerdes gives examples from Ancient Egypt, the Tellem culture in West Africa, Somalia, Angola and Senegal. The toothed squares embody interesting mathematical information. Professor Gerdes demonstrates how interesting formulae can be derived from these squares. Moreover, he shows that some of the concepts associated with the Pythagorean Theorem are embodied in these squares.

Counting the rows of black squares, starting from the top to the bottom, there is one square in the top row, two in the second, three in the third, four in the fourth, three in the fifth, two in the sixth and one in the seventh. Since there are sixteen black squares in total, arranged in a four by four arrangement, tilted by 45 degrees, this shows that $4^2 = 1 + 2 + 3 + 4 + 3 + 2 + 1$.

This equation could be rearranged be subtracting 4 from both sides to arrive at $4^2 - 4 = (1 + 2 + 3) + (3 + 2 + 1) = 2 \times (1 + 2 + 3)$. This formula could be rearranged to arrive at $(4^2 - 4)/2 = 1 + 2 + 3$. Turning this into a general formula gives $(n^2 - n)/2 = 1 + 2 + 3 ... (n - 1)$

Counting the total number of black squares, there are 16 arranged in a four by four pattern, tilted by 45 degrees. Counting the white squares, there are 9 arranged in a three by three pattern, tilted by 45 degrees. The total number of squares is 25. This shows that $(3 \times 3) + (4 \times 4) = 25$. Turning this into a general formula gives $T = a^2 + b^2$, where T is the area of a toothed square, a is the number of black squares, and b is the number of white squares.

Figure 42. Toothed square designs exist in many parts of Africa. They are found in 18th Dynasty Egypt, the 11th century Tellem culture in West Africa. They are also known in Somalia, Angola and Senegal.

What does this prove? Professor Gerdes suggests that the designers of the toothed squares could have derived the Pythagorean Theorem from it. Professor Arthur Powell makes the stronger claim that: "Is it not possible, then, to claim African originality of these mathematical ideas? Clearly, cultural groups in other geographical ... regions do and can make similar claims". However, I take the much more cautious view that the designers of the toothed square were clearly familiar with some of the concepts that we have come to associate with the Pythagorean Theorem. However, we would need to see evidence that they actually related their knowledge of these concepts to triangles before we could definitively credit them with the full knowledge of the Pythagorean Theorem.

From Central African Mats to Latin Squares

Professor Gerdes shows that some of the Tchokwe and Kuba mat makers produced mats that combined two colour patterns in a particularly interesting way. Apparently, their mats solved the problem of weaving vertical and horizontal strands of two different colours over and under to create an overall mat of a single colour covered by a square grid of dots of the second colour. Each vertical dark strand passed over one horizontal white strand and under four horizontal white strands. Each horizontal white strand passed under one vertical dark strand and over four vertical dark strands. The pattern repeats itself after 5 strands in both directions. Gerdes suggests that the pattern may be considered as a series of 5 x 5 blocks.

The result could be shown on squared paper. It could also be called the (1, -2) solution since from every dark dot to the next one, we move one step across and two steps down.

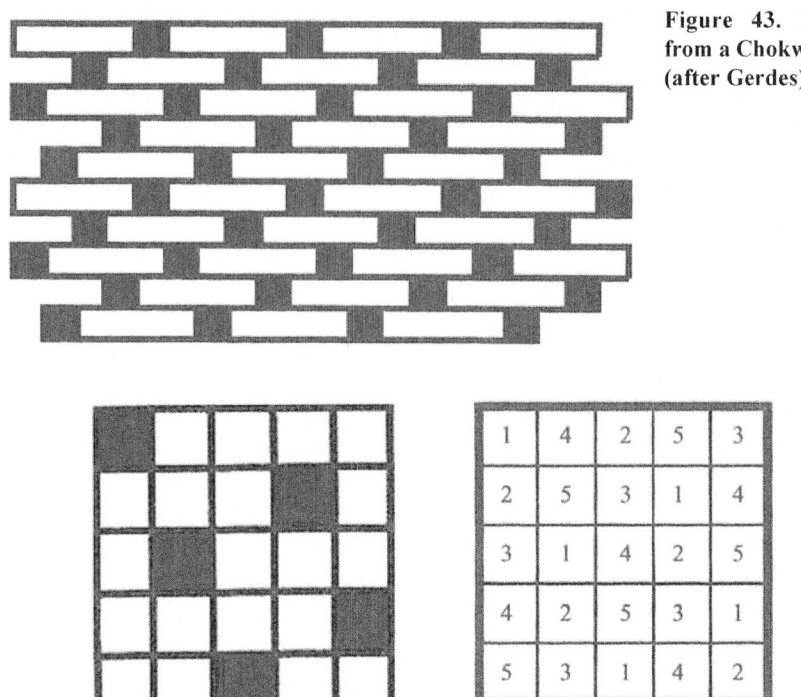

Figure 43. Detail from a Chokwe mat (after Gerdes).

1	4	2	5	3
2	5	3	1	4
3	1	4	2	5
4	2	5	3	1
5	3	1	4	2

Figure 44a. Presenting the Tchokwe mat pattern on squared paper (after Gerdes).
Figure 44b. Derivation of a Latin Square from the design.

Gerdes shows that once we start to number the dark squares 1 to 5, we could also give the white squares next to them unique numbers to construct a Latin square. In each row or column the numbers 1, 2, 3, 4 and 5 appear just once. I argue that it is simpler to give all of the dark squares the same number (perhaps '1'). We could give the white square below it another unique number (perhaps '2'). We could give '3' to the next white square, etcetera. I agree with Gerdes that one arrives at a Latin square but our methodologies are slightly different. Thus it is reasonable to conclude that the thinking behind the Tchokwe and Kuba mat making was equivalent to the thinking behind a Latin square.

CONCLUSION

Summing Up

Africa has suffered greatly over the last five hundred years and this has caused great archival discontinuity and disrupted its mathematical development. Some of the Africans who were captured and enslaved in the United States continued to show the high intellectual calibre representative of the Africa from which they came. Thomas Fuller, for example, was an African shipped to America in 1724. He had remarkable powers of calculation, and late in his life was discovered by antislavery campaigners who used him to prove that Blacks are not mentally inferior to whites. Born somewhere between present day Liberia and Benin, he was enslaved by Elizabeth Cox of Alexandria in Virginia.

He could multiply 9 digit numbers as mental arithmetic. A contemporaneous source says:

> He could give the number of months, days, weeks, hours, minutes, and seconds in any period of time that any person chose to mention, allowing in his calculation for all leap years that happened in the time; he would give the number of poles, yards, feet, inches, and barley-corns in any distance, say the diameter of the earth's orbit; and in every calculation he would produce the true answer in less time than ninety-nine men out of a hundred would produce with their pens.

Documents at the time record that:

> When he was about seventy years old, two gentlemen, natives of Pennsylvania, viz., William Hartshorne and Samuel Coates, men of probity and respectable characters, having heard, in travelling through the neighbourhood in which the slave lived, of his extraordinary powers in arithmetic, sent for him and had their curiosity sufficiently gratified by the answers which he gave to the following questions: First, upon being asked how many seconds there were in a year and a half, he answered in about two minutes, 47 304 000. Second: On being asked how many seconds a man has lived who is 70 years, 17 days and 12 hours old, he answered in a minute and a half 2 210 500 800. One of the gentlemen who employed himself with his pen in making these calculations told him he was wrong, and the sum was not so great as he had said - upon which the old man hastily replied: stop, master, you forget the leap year. On adding the amount of

the seconds of the leap years the amount of the whole in both their sums agreed exactly.

Connor and Robertson, two authorities on mathematics history, claim: "Present day thinking is that Fuller learnt to calculate in Africa before he was brought to the United States as a slave."

Where the scholarship needs to go

Despite having some information to discuss, presented in this book, our knowledge of African mathematical history remains very limited indeed. I will use this section of the book to outline some suggestions for where African mathematical research needs to go.

Professor Diop reports that the German archaeologist Lepsius discovered in the Sudanese city of Meroë the foundation of an astronomical observatory. On the walls of the edifice was found a scene representing people operating an instrument resembling an astrolabe. He also found a series of numerical equations relating to astronomic events which occurred two centuries BC. What are these equations? Who has published them?

Professor Chancellor Williams, the great African American scholar, mentions the same Sudanese writing system and says that they developed a numerical system for mathematics. This again raises questions. Are there Kushite mathematical documents? Has anyone published them? What do they say?

Thousands of documents from Mediaeval Nubia have survived at the site of Qasr Ibrim. A quarter of a million manuscripts have survived in Ethiopia. Professor Albert Churchward writing in 1910 mentions evidence that one of the southern African ruins associated with the Empire of Munhumutapa contained manuscripts covered with writing. The questions these manuscript collections raise are the same as before: What are the mathematical content (if any) of these manuscripts? Only the research of Professor Neugebauer seems to have addressed any of these questions.

Then we have the fifteenth century Timbuktu manuscripts of which Professor Medupe spoke of when he said:

> Other manuscripts dating back 600 years include beautifully drawn diagrams of the orbits of planets, which demonstrate the use of complex mathematical calculations.

What are these calculations? Has anyone published them? There are perhaps one million old manuscripts that have survived across West Africa.

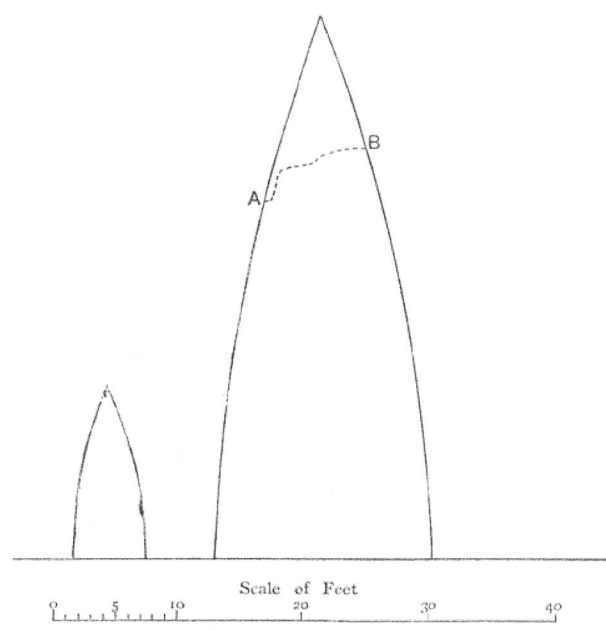

Figure 45. Partially restored diagrammatic sketch of the two towers at Great Zimbabwe from J. Theodore Bent's *The Ruined Cities of Mashonaland*, p.149. They appear as F and D on the plan, see next page.

Scale of Feet

Many are known to have covered scientific subjects. Professor Paulus Gerdes made an important comment on Timbuktu mathematics:

> Hundreds of mathematical manuscripts--written in Arabic and in various African languages--from Timbuktu in today's Mali remain to be analysed ...

Finally, there is the claim by Robert M. W. Swan, a mining engineer, who wrote a controversial 1891 paper entitled *On the Orientation and Measurements of Zimbabwe Ruins*. Mr J. Theodore Bent included this paper as Chapter V of his *The Ruined Cities of Mashonaland*.

In the paper, Swan presented measurements and calculations that implied that π was built into the very structure of the Great Zimbabwe Temple, an ancient or mediaeval Southern African architectural structure. Scholars have since rubbished these claims but since 1891 it is undeniable that parts of the Temple have been demolished and rebuilt destroying the mathematical relationships that Swan detected in 1891. His claim is that the circumference of one of the cone towers (i.e. F) is the same as the diameter of the great cone tower (i.e. D) showing a π relationship between the two. Are there any methods of checking the pre 1891 dimensions of the Temple to confirm or disconfirm Swan's claims?

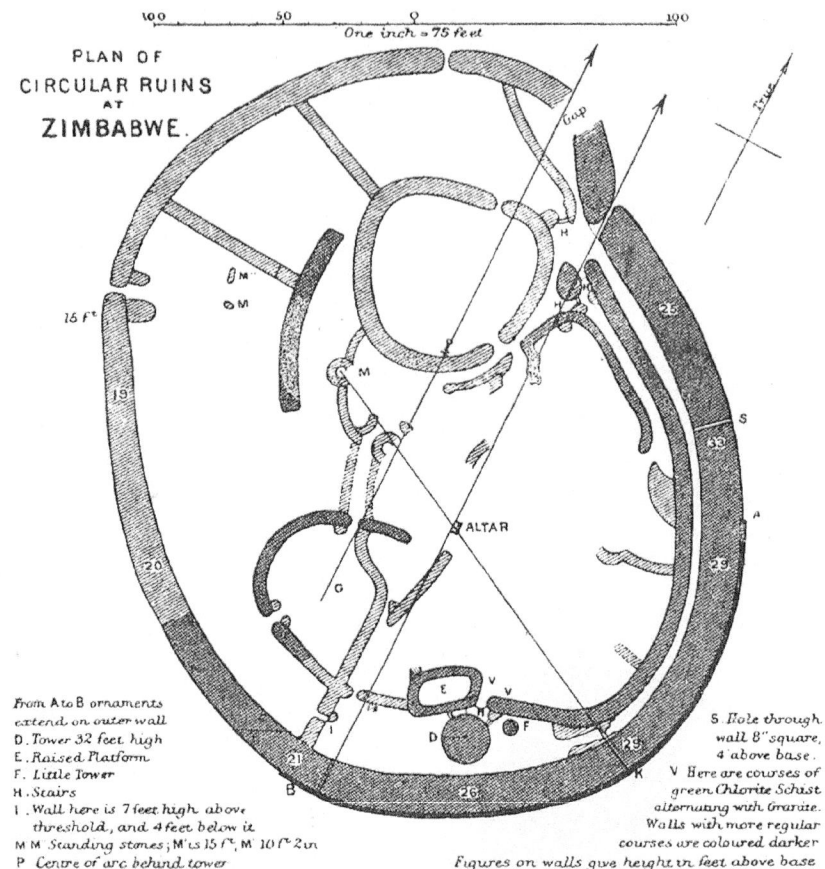

Figure 46. Plan of the Great Zimbabwe Temple from J. Theodore Bent's *The Ruined Cities of Mashonaland*, opposite p.104. The two towers are D and F on the plan. It was claimed that the circumference of F was equal to the diameter of D.

BIBLIOGRAPHY

Introduction

Ibrahima Diallo Sambegou of Guinea solves 270-year-old math problem, 3 April 2013, see internet at http://hobnobdrive.com/culture/ibrahima-diallo-sambegou-of-guinea-solves-270-year-old-math-problem/

Best Masire, *Great Mathematician in Guinea,* 4 December 2013, see internet at http://newsofthesouth.com/great-mathematician-in-guinea/

See also http://wadr.org/fr/site/news_fr/4068/Un-guin%C3%A9en-solutionne-un-probl%C3%A8me-de-math-vieux-de-270-ans.htm

Mamokgethi Setati & Abdul Karim Bangura, *African Mathematics: From Bones to Computers,* US, University Press of America, 2011, pp.21-22

Chapter 1: African Proto-Mathematics

CNN, *The Ishango Bone,* see internet at https://www.youtube.com/watch?v=VX90lN8VSME

Charles Finch, *The Star of Deep Beginnings,* US, Khenti, 1998, pp.55-57

Franz Gnaedinger, *Very Early Calendars,* 2005-2008, see internet at http://www.seshat.ch/home/calendar.htm

José Naranjo, *The Ishango Bone,* 27 July 2010, see internet at http://www.guinguinbali.com/index.php?lang=en&mod=news&task=view _news&cat=2&id=708

Mamokgethi Setati & Abdul Karim Bangura, *African Mathematics: From Bones to Computers,* US, University Press of America, 2011, pp.11-15

Ta Neter Foundation, *Ancient African Mathematics,* 2011, see internet at http://taneter.org/math.html

Chapter 2: Ancient Egyptian Mathematics

Charles Finch, *The Star of Deep Beginnings,* US, Khenti, 1998, pp.57-61

Beatrice Lumpkin, *African & African-American Contributions to Mathematics,* US, Portland Public Schools, 1987, pp.40-41

Théophile Obenga, *African Philosophy: The Pharaonic Period: 2780-330 BC,* Senegal, Per Ankh, 2004, pp.421, 443-448, 495-498

Mamokgethi Setati & Abdul Karim Bangura, *African Mathematics: From Bones to Computers,* US, University Press of America, 2011, pp.44-45

Chapter 3: The Rhind Mathematical Papyrus

Cheikh Anta Diop, *Civilization or Barbarism,* US, Lawrence Hill Books, 1991, pp.238-241, 260-274

Charles Finch, *The Star of Deep Beginnings,* US, Khenti, 1998, pp.78-88

Beatrice Lumpkin, *African & African-American Contributions to Mathematics,* US, Portland Public Schools, 1987, pp.20-25

Théophile Obenga, *African Philosophy: The Pharaonic Period: 2780-330 BC,* Senegal, Per Ankh, 2004, pp.421-431, 437-441, 443-451, 475-480, 485-487, 490-498

Chapter 4: Other Egyptian Papyri

Cheikh Anta Diop, *Civilization or Barbarism,* US, Lawrence Hill Books, 1991, pp.232-237, 251-258, 267, 269, 273-274

Charles Finch, *The Star of Deep Beginnings,* US, Khenti, 1998, pp.83-88

Beatrice Lumpkin, *African & African-American Contributions to Mathematics,* US, Portland Public Schools, 1987, p.26

Théophile Obenga, *African Philosophy: The Pharaonic Period: 2780-330 BC,* Senegal, Per Ankh, 2004, pp.439-440, 481-484, 487-489

Chapter 5: Egyptian Mathematical Evidence from non-documentary sources

Robert Bauval and Graham Hancock, *Keeper of Genesis,* UK, Mandarin, 1996, pp.38-44

Cheikh Anta Diop, *Civilization or Barbarism,* US, Lawrence Hill Books, 1991, pp.248-251, 258-260, 272

Charles Finch, *The Star of Deep Beginnings,* US, Khenti, 1998, pp.71-75, 89-91, 117

Beatrice Lumpkin, *Mathematics and Engineering in the Nile Valley* from *Egypt: Child of Africa,* edited by Ivan Van Sertima, US, Transaction Publishers, 1994, p.332

Peter Tompkins, *Secrets of The Great Pyramid,* US, Harper & Row, 1971, pp.189-200

Chapter 6: Mathematics in the Time of the Moors

Captions from *Exact Sciences, Parts 1.1, 1.2, 1.3, 1.4* in The Science Museum, Granada

Paulus Gerdes and Ahmed Djebbar, *Mathematics in African History and Cultures: 2nd Edition,* US, Lulu, 2007, pp.16-17

Karen Carr, *African Mathematics,* 2012-2014, see internet at http://www.historyforkids.org/learn/africa/science/numbers.htm

Ibn al-Yasamin al-Ishbilli, see internet at http://www.muslimheritage.com/scholars/ibn-al-yasamin-al-ishbilli

IBNU AL-YASAMIN AL-ISHBILLI, in *REPUBLIKA khanaza,* 4 August 2010, see internet at http://mirror.unpad.ac.id/koran/republika/2010-08-04/republika_2010-08-04_028.pdf

Friedrich Katscher, *Extracting Square Roots Made Easy: A Little Known Medieval Method,* 2014, see internet at http://www.maa.org/publications/periodicals/convergence/extracting-square-roots-made-easy-a-little-known-medieval-method-historical-background

J. J. O'Connor and E. F. Robertson, *al-Marrakushi ibn Al-Banna*, 1999, see internet at
http://www-history.mcs.standrews.ac.uk/Mathematicians/Al-Banna.html

J. J. O'Connor and E. F. Robertson, *Muhyi l'din al-Maghribi*, 1999, see internet at http://www-history.mcs.st-andrews.ac.uk/Biographies/Al-Maghribi.html

Mamokgethi Setati & Abdul Karim Bangura, *African Mathematics: From Bones to Computers*, US, University Press of America, 2011, pp.50-59

Chapter 7: Mathematics in Ethiopia

Otto Neugebauer, *Chronography in Ethiopic Sources*, Germany, Osterreiche Akademie Der Wissenschaften, 1989, pp.27-30

Otto Neugebauer, *Ethiopic Astronomy and Computus*, Germany, Osterreiche Akademie Der Wissenschaften, 1979, pp.7-10, 13-17, 18, 27-66, 111, 123-124, 132, 175-177, 221-222

David Okuefuna (Executive Producer), *Go Forth and Multiply*, UK, The Open University for BBC 4, Television Programme, 2005

Chapter 8: Shared Games and Mathematical Concepts in Ethiopia and the Rest of Africa

Alicia, *Cornrows and Fractals*, 14 July 2011, see internet at http://ethiopia.limbo13.com/index.php/cornrows_and_fractals/

Ron Eglash, *African Fractals*, US, Rutgers University Press, 1999, pp.20-33

Ron Eglash, *The fractals at the heart of African designs*, 2007, see internet at http://www.ted.com/talks/ron_eglash_on_african_fractals/transcript

Mamokgethi Setati & Abdul Karim Bangura, *African Mathematics: From Bones to Computers*, US, University Press of America, 2011, pp.5, 75-82

Ta Neter Foundation, *Ancient African Mathematics*, 2011, see internet at http://taneter.org/math.html

Claudia Zaslavsky, *Africa Counts*, US, Lawrence Hill, 1973, pp.116-136

Chapter 9: Mathematics in West and Central Africa

Ron Eglash, *African Fractals*, US, Rutgers University Press, 1999, pp.20-33

Charles S. Finch, *The Star of Deep Beginnings*, US, Khenti, 1998, pp.91-92

Paulus Gerdes, *Drawings from Angola: Living Mathematics*, US, Lulu, 2007, pp.7-9, 17-30, 35-48

Paulus Gerdes, *Geometry From Africa: Mathematical and Educational Explorations*, US, The Mathematical Association of America, 1997, pp.157-176

Paulus Gerdes and Ahmed Djebbar, *Mathematics in African History and Cultures: 2nd Edition*, US, Lulu, 2007, pp.73-75, 105-107, 208

Dirk Huylebrouck, *Mathematics in (central) Africa before colonization*, in Anthropologica et Præhistorica, 117, 2006, pp.143-144

Beatrice Lumpkin, *African & African-American Contributions to Mathematics*, US, Portland Public Schools, 1987, pp.40-41

Théophile Obenga, *African Philosophy: The Pharaonic Period: 2780-330 BC*, Senegal, Per Ankh, 2004, pp.470-474

Mamokgethi Setati & Abdul Karim Bangura, *African Mathematics: From Bones to Computers*, US, University Press of America, 2011, pp.16-38

Claudia Zaslavsky, *Africa Counts*, US, Lawrence Hill, 1973, pp.39-46, 103-109, 172-193, 202-207, 213-220

Chapter 10: Mathematics in the West African superstates

Curtis Abraham, *Stars of the Sahara,* in *New Scientist,* Issue 2617, 15 August 2007, pp.39-41

Ron Eglash, *African Fractals,* US, Rutgers University Press, 1999, pp.20-33

Charles S. Finch, *The Star of Deep Beginnings,* US, Khenti, 1998, pp.93-94

Henry Louis Gates, *Into Africa,* Television Series Part 5, *The Road To Timbuktu,* UK, BBC Television, 1999

Paulus Gerdes, *Geometry From Africa: Mathematical and Educational Explorations,* US, The Mathematical Association of America, 1999, pp.6-31

Paulus Gerdes and Ahmed Djebbar, *Mathematics in African History and Cultures: 2nd Edition,* US, Lulu, 2007, pp.183, 212

John O. Hunwick & Alida Jay Boye, *The Hidden Treasures of Timbuktu: Historic City of Islamic Africa,* UK, Thames & Hudson, 2008, p.90

Ahmad Kani, *Arithmetic in the pre-colonial Central Sudan,* 1992, see internet at http://www.africahistory.net/kani.htm

Karen E. Lange, *Djénné: West Africa's Eternal City,* in *National Geographic,* US, June 2001, p.110

Théophile Obenga, *African Philosophy: The Pharaonic Period: 2780-330 BC,* Senegal, Per Ankh, 2004, pp.489-490

Mamokgethi Setati & Abdul Karim Bangura, *African Mathematics: From Bones to Computers,* US, University Press of America, 2011, pp.18-19

Claudia Zaslavsky, *Africa Counts,* US, Lawrence Hill, 1973, pp.137-151, 276

Chapter 11: Mathematical ideas in traditional African know-how, crafts and designs

Paulus Gerdes, *Geometry From Africa: Mathematical and Educational Explorations,* US, The Mathematical Association of America, 1999, pp.vi, 9, 10, 36, 37, 38, 69-80, 94-96, 110-121

Conclusion

Curtis Abraham, *Stars of the Sahara,* in *New Scientist,* Issue 2617, 15 August 2007, pp.39-41

J. Theodore Bent, *The Ruined Cities of Mashonaland,* UK, Longmans, Green and Co., 1902, pp.104, 141-178

Albert Churchward, *The Signs and Symbols of Primordial Man,* UK, George Allen & Co., 1910, p.75

Cheikh Anta Diop, *Precolonial Black Africa,* US, Lawrence Hill Books, 1987, pp.196-198

J. J. O'Connor and E. F. Robertson, *Thomas Fuller,* 2005, see internet at http://www-history.mcs.st-and.ac.uk/Biographies/Fuller_Thomas.html

Laurance R. Doyle, *Astronomy of Africa,* in *Encyclopaedia of the History of Science, Technology and Medicine in Non-Western Cultures,* see internet at http://www.safaris.cc/8art.encyclo.htm

Paulus Gerdes, *Sona Geometry From Angola,* Italy, Polimetrica, 2006, p.217

Derek A. Welsby, *The Medieval Kingdoms of Nubia,* UK, The British Museum Press, 2002, p.24

Chancellor Williams, *The Destruction of Black Civilization,* US, Third World Press, 1987, p.127

Dr Scott W. Williams, *Thomas Fuller: African slave and mathematician 1710-1790,* see internet at http://www.math.buffalo.edu/mad/special/fuller_thomas_1710-1790.html

PART TWO

LEARN AFRICAN MATHEMATICS IN 21 LESSONS

INTRODUCTION

This part of the book gives teachers and learners the opportunity to engage with the mathematical ideas in the history section. The level of difficulty is pitched at around Year 7 level. In other words, this is the level of material that a pupil in the first year of Secondary School might encounter.

We have produced material for 21 one-classes. Each class refers to an idea or theme in Part One of the book. What follows are questions for the pupils to attempt, discuss and critique. The order of the classes follows the order the material was discussed in Part One of the book. This is not always ideal since harder mathematical topics appear in the book before some of the easier topics. Any teacher or student should bear this in mind and re-arrange the order of the classes as necessary.

LESSON 1: IBRAHIMA AND GOLDBACH'S CONJECTURE

Introduction

Ibrahima Diallo Sambegou of Guinea produced a solution to an old maths problem called Goldbach's Conjecture. Goldbach believed that every even number greater than two can be written as the sum of two prime numbers.

Ibrahima explains: "For illustration, we see that $6 = 3 + 3$, $8 = 3 + 5$, $3 + 7 = 10$ or $5 + 5$, $30 = 11 + 19 = 13 + 17$, $100 = 17 + 83$ ... is it true [of] any even number? This is the clue." The numbers that he mentions in building these equations--3, 5, 7, 11, 13, 17, 83--are all prime numbers, i.e. only divisible by themselves or by one to produce a whole number.

Questions

Find two prime numbers in each case that add up to the following numbers:

1. 4	**2.** 12	**3.** 14	**4.** 16	**5.** 18
6. 20	**7.** 22	**8.** 24	**9.** 26	**10.** 28
11. 32	**12.** 34	**13.** 38	**14.** 42	**15.** 46
16. 50	**17.** 54	**18.** 58	**19.** 62	**20.** 66
21. 70	**22.** 74	**23.** 78	**24.** 82	**25.** 86
26. 90	**27.** 94	**28.** 98	**29.** 102	**30.** 106

LESSON 2: THE LEBOMBO BONE

Introduction

The Lebombo Bone is a 37,000 year old artefact with 29 notches carved into it. The 29 notches are thought to represent a lunar calendar. There are 30 spaces between and next to the 29 notches. One should read the spaces and notches as follows: 30 spaces plus 29 notches plus 30 spaces plus 29 notches plus 30 spaces, etcetera, yields 30 29 30 29 30 ... nights. Moreover, the information could just as easily yield 30 59 89 118 148 177 207 236 266 295 325 354 nights for 1 2 3 4 5 6 7 8 9 10 11 12 lunations.

A lunation is the average time of one lunar phase cycle. The length of any one lunar month can vary from 29.26 to 29.8 days. Most writers present it as approximately 29.5 days. Thus the Bone allows us to approximately calculate one lunation as 30 nights, two lunations as 59 nights (i.e. $30 + 29$), three lunations as 89 nights (i.e. $30 + 29 + 30$), etcetera.

Questions

Work out the following questions without a calculator.
1. How many nights occur in 3 lunations?
2. How many nights occur in 8 lunations?
3. How many nights occur in 7 lunations?
4. How many nights occur in 5 lunations?
5. How many nights occur in 1 lunar year and 6 lunations?
6. How many nights occur in 2 lunar years and 3 lunations?
7. How many nights occur in 4 lunar years?
8. How many nights occur in 5 lunar years and 4 lunations?
9. How many nights occur in 2 lunar years and 8 lunations?
10. How many nights occur in 6 lunar years and 8 lunations?
11. How many lunar months have gone after 708 nights?
12. How many lunar months have gone after 413 nights?
13. How many lunar months have gone after 531 nights?
14. How many lunar years and months have gone after 1180 nights?
15. If a lunation cycle began on a month which had 30 days, how many nights will occur during the next 9 lunar months?
16. If a lunation cycle began on a month which had 30 days, how many nights will occur during the next 13 months?

17. If a lunation cycle begins on a month which has 29 days, how many nights will occur during the next lunar year and 4 lunar months?

18. If a lunation cycle begins on a month which has 30 days, how many nights will occur during the next 6 lunar years?

19. How many lunar months have gone after 4248 nights?

20. How many lunar months have gone after 1534 nights?

LESSON 3: THE ISHANGO BONE

Introduction

The Ishango Bone has a number pattern of 3, 6, 4, 8, 10, 5, 5, 7 which is interpreted as:

(i) 3 + 6 = 9 which is 12 - 3
(ii) 4 + 8 = 12
(iii) 10 + 5 = 15 which is 12 + 3
(iv) Finally, 5 + 7 = 12.

The number 12 is clearly the central number. Notice that the third and fourth numbers always adds up to the central number. Notice also that the seventh and eighth number also adds up to the central number.

Questions

1. Suppose a bone was newly discovered with the numbers 4, 8, 8, 8, 17, 3, 5, 11. Use the above pattern to show that the central number is 16.

2. Suppose the bone had the numbers 7, 7, 10, 11, 21, 7, 9, 12. Use the above pattern to show that the central number is 21.

3. If a bone had the numbers 8, 16, 12, 24, 40, 8, 5, 31, show that the central number is 36.

4. If a bone had the numbers 1, 4, 6, 4, 10, 5, 5, 5, show that the central number is 10.

5. If a bone had the numbers 2, 8, 13, 7, 20, 10, 8, 12, show that the central number is 20.

6. If a bone had the numbers 9, 16, 20, 30, 39, 36, 5, 45, suggest what the central number might be.

7. If a bone had the numbers 6, 29, 28, 14, 42, 7, 30, 12, suggest what the central number might be.

8. If a bone had the numbers 17, 39, 24, 40, 49, 23, 5, 59, suggest what the central number might be.

9. If a bone had the numbers 30, 60, 40, 80, 100, 50, 108, 12, suggest what the central number might be.

10. If a bone had the numbers 23, 85, 60, 84, 97, 73, 5, 139, suggest what the central number might be.

LESSON 4: EGYPTIAN NUMERALS

Introduction

The Egyptians wrote their numbers using hieroglyphic symbols.

Questions

Decode the following numbers. The Egyptians often read from right to left!

1. ⎮∩∩∩𝟡𝟡𝟡⎮⎮⎮⎮↘

2. ⎮⎮⎮𝟡𝟡𝟡↥↥↘

3. ⎮⎮⎮∩∩⎮↘↘↘⚚ ⎮⎮⎮∩

4. ⎮⎮𝟡↥⎮⚚

5. ⎮⎮⎮⎮∩∩∩↥↥↥↘⎮↘ ⎮⎮⎮∩∩ ↥↥↥

6. ∩∩∩𝟡↥↘

7. ∩∩↘↘↘⚚⚚⚚⚚⚚

8. 𝟡⎮⎮⎮⎮

9. ↥⎮⎮⎮⚚

10. ⎮⚚⚚

11. Write 23 using hieroglyphs
12. Write 56 using hieroglyphs
13. Write 2023 using hieroglyphs
14. Write 6967 using hieroglyphs
15. Write124598 using hieroglyphs
16. Write 406 using hieroglyphs
17. Write 3090021 using hieroglyphs
18. Write 72593 using hieroglyphs
19. Write 8888 using hieroglyphs
20. Write 9999 using hieroglyphs

Discussion Question

With reference to your answers to questions 19 and 20, why do you think the Ancient Egyptians invented simpler number systems to replace hieroglyphics?

LESSON 5: UNIT FRACTIONS

Introduction

The Egyptians had fractions given as 1/2, 2/3, 1/3, 1/4, 1/5, 1/6, etcetera. With the exception of 2/3, all the other fractions were expressed as one over two, one over three, and so on. These are called unit fractions. If the Ancient Egyptians wanted to express 3/4, they would write 1/2 + 1/4. In a similar vein, if they wanted to express 2/5, they would write 1/3 + 1/15.

Questions

In each case (a) What are these unit fractions as a single fraction? (b) What are they in decimals?
   1. 1/2 + 1/4 + 1/8
   2. 1/4 + 1/8 + 1/16
   3. 1/3 + 1/6
   4. 1/3 + 1/6 + 1/12
   5. 1/3 + 1/6 + 1/12 + 1/24
   6. 1/4 + 1/16 + 1/32
   7. 1/2 + 1/4 + 1/8 + 1/16 + 1/32 + 1/64
   8. 1/5 + 1/10 + 1/20
   9. 1/5 + 1/10 + 1/30
   10. 1/5 + 1/15 + 1/45
   11. 1/5 + 1/10 + 1/20 + 1/40 + 1/80
   12. 1/7 + 1/14 + 1/28
   13. 1/3 + 1/4 + 1/6 + 1/8 + 1/24
   14. 1/6 + 1/8 + 1/12 + 1/16 + 1/48
   15. 1/5 + 1/6 + 1/10 + 1/15 + 1/30
   16. 1/10 + 1/12 + 1/20 + 1/30 + 1/60
   17. 1/5 + 1/7 + 1/35
   18. 1/7+ 1/10 + 1/14 + 1/70
   19. 1/7 + 1/21 + 1/63
   20. 1/2 + 1/14 + 1/42 + 1/126

LESSON 6: ANCIENT EGYPTIAN MULTIPLICATION

Introduction

When multiplying two numbers together, the Egyptians would make one number the multiplier and the other number the multiplicand. Suppose they wanted to multiply five by seventeen, five would be the multiplier and seventeen would be the multiplicand.

Always starting from number one, they would keep doubling the number until the number five was reached. They would also double the multiplicand starting with the number seventeen.

Since they were multiplying by five, the Egyptians would use the information concerning the multiplier of 1 (and the corresponding 17) and the multiplier of 4 (and the corresponding 68). They chose these figures because $1 + 4 = 5$. Therefore the calculation would be $17 + 68 = 85$. They would disregard the 2 and the corresponding 34.

Multiplier	Multiplicand
1 Use this	17 Use
2 Reject this	34 Reject this
4 Use this	68 Use
$1 + 4 = 5$	$17 + 68 = 85$

Questions

Using this method, multiply the following numbers together. Make the first number the multiplier. Make the second number the multiplicand.

1. 3 x 4
2. 4 x 6
3. 9 x 13
4. 13 x 16
5. 17 x 31
6. 19 x 36
7. 23 x 68
8. 47 x 149
9. 49 x 158.2
10. 51.25 x 2002

Discussion Questions

1. (a) Did you know the 13 times table before attempting these questions?
(b) Did you know the 17 times table before attempting these questions?
(c) Did you know the 19 times table before attempting these questions?
 2. Could you do question 10?
3. Explain the advantages and disadvantages of using the Ancient Egyptian method.

Questions

The following questions are the same as before and should produce the same answer except that the order of the numbers has been reversed. Again make the first number the multiplier and the second number the multiplicand.

1. 4 x 3
2. 6 x 4
3. 13 x 9
4. 16 x 13
5. 31 x 17
6. 36 x 19
7. 68 x 23
8. 149 x 47
9. 158.2 x 49
10. 2002 x 51.25

Discussion Questions

1. Could you do question 9?
2. What difficulty did you experience in doing question 10?
3. Explain other advantages and disadvantages of using the Ancient Egyptian method.

LESSON 7: ANCIENT EGYPTIAN DIVISION

Introduction

Suppose the Egyptians wanted to divide 425 by 18 they would arrange the numbers into two columns looking like this:

1	18
2	36
4	72
8	144
16	288
32	576

Since 576 is larger than 425 the Egyptians would use the number 16 (and the corresponding 288). Now from 425 subtract the partial products, starting from the number 288. The results would look like this:

```
  425
- 288      (16 x 18)
  137
-  72      (4 x 18)
   65
-  36      (2 x 18)
   29
-  18      (1 x 18)
   11      remainder
```

The answer is 16 + 4 + 2 + 1 = 23. The remainder is 11.

Questions

Using this method, divide the first number by the second number. Make the second number the starting point for the doubling in the right hand column. Make the number 1 the starting point for the doubling in the left hand column.

1. 6 ÷ 2
2. 12 ÷ 3
3. 36 ÷ 4

4. $63 \div 7$
5. $96 \div 12$
6. $143 \div 14$
7. $287 \div 15$
8. $374 \div 17$
9. $390 \div 18$
10. $425 \div 19$

LESSON 8: THE CYLINDER

Introduction

The RMP Problems 41, 42 and 43 address the volumes of cylindrical shaped granaries. Problem 41, for example, asks: "Find the volume of a cylindrical granary of diameter 9 and height 10." The formula used was Volume = d x (8/9)² x h

Thus 9 x 8/9 = 8
8² = 64
64 x 100 = 640

Questions

Calculate the volumes of the following cylinders using the RMP formula. Do not use a calculator. Make sure that your answers are in mm³, cm³ or m³.
   1. Cylinder with a radius of 9 mm and a height of 10 mm
   2. Cylinder with a diameter of 63 cm and a height of 20 cm
   3. Cylinder with a radius of 13.5 m and a height of 5 m
   4. Cylinder with a diameter of 45 mm and a height of 40 mm
   5. Cylinder with a radius of 40.5 m and a height of 80 m
   6. Cylinder with a diameter of 108 cm and a height of 200 cm
   7. Cylinder with a radius of 49.5 m and a height of 180 m
   8. Cylinder with a diameter of 36 mm and a height of 50 mm
   9. Cylinder with a radius of 36 m and a height of 90 m
   10. Cylinder with a diameter of 54 cm and a height of 60 cm

Notes

The RMP formula implies a value for π of 256/81 = 3.16049. Mathematicians since then have come up with a more accurate figure for π. To five decimal places it is 3.14159. The RMP Problem 41 using the more modern formula is Area = π x r² x h

Thus 3.14159 x 4.5² x 10 = 63.6171975 x 10 = 636.171975

636.171975 is a more accurate figure than the RMP answer of 640.

Questions

Using this formula, evaluate the questions below using a calculator. They are the same questions as before. As before make sure that your answers are in mm³, cm³ or m³.

1. Cylinder with a radius of 9 mm and a height of 10 mm

2. Cylinder with a diameter of 63 cm and a height of 20 cm

3. Cylinder with a radius of 13.5 m and a height of 5 m

4. Cylinder with a diameter of 45 mm and a height of 40 mm

5. Cylinder with a radius of 40.5 m and a height of 80 m

6. Cylinder with a diameter of 108 cm and a height of 200 cm

7. Cylinder with a radius of 49.5 m and a height of 180 m

8. Cylinder with a diameter of 36 mm and a height of 50 mm

9. Cylinder with a radius of 36 m and a height of 90 m

10. Cylinder with a diameter of 54 cm and a height of 60 cm

LESSON 9: CUBING A NUMBER

Introduction

Cubing a number is where a number is multiplied by itself and then multiplied by itself a second time. In measurement, a number 'to the power of one' or 10^1 has not been multiplied by anything and represents length. In this example, the length is 10 units long. In measurement, a number 'to the power of two' or 10^2 meaning multiplied by itself represents the area of a square. In this example, the area is 100 units squared. Finally, in measurement, a number 'to the power of three' or 10^3 represents the volume of a cube. In this example, the volume is 1000 units cubed. Thus 10^3 is 10 x 10 x 10 or as a formula $V = a \times a \times a$, where V is the volume and a is the length of each side of the cube.

Questions

Cube the following numbers without using a calculator:

1. 1	**2.** 2	**3.** 4	**4.** 5	**5.** 8
6. 11	**7.** 14	**8.** 17	**9.** 21	**10.** 26
11. 31	**12.** 37	**13.** 44	**14.** 52	**15.** 61
16. 71	**17.** 82	**18.** 94	**19.** 107	**20.** 121

LESSON 10: AREA OF A CIRCLE

Introduction

The RMP Problem 50 addresses the area of a circle. The question reads as follows: "Example of a round field of diameter 9 khet; what is the area?" The formula used was Area = d x $(8/9)^2$.

Thus 9 x 8/9 = 8
8^2 = 64

Questions

Calculate the areas of the following circles using the RMP formula. Do not use a calculator. You must convert all of the lengths into diameters before making the calculation. Make sure that your answers are in mm^2, cm^2 or m^2.
1. Circle with a diameter of 18 cm
2. Circle with a radius of 31.5 m
3. Circle with a diameter of 27 mm
4. Circle with a radius of 22.5 m
5. Circle with a diameter of 81 cm
6. Circle with a radius of 54 m
7. Circle with a diameter of 99 mm
8. Circle with a radius of 18 m
9. Circle with a diameter of 72 cm
10. Circle with a radius of 27 m

Notes

The RMP formula implies a value for π of 256/81 = 3.16049. Mathematicians since then have come up with a more accurate figure for π. To five decimal places it is 3.14159. The RMP Problem 50 using the more modern formula is Area = π x r^2.

Thus 3.14159 x 4.5^2 = 63.6171975.

63.6171975 is a more accurate figure than the RMP answer of 64.

Questions

Using this formula, evaluate the questions below using a calculator. They are the same questions as before but you must convert all of the lengths into radii before making the calculation. As before, make sure that your answers are in mm^2, cm^2 or m^2.

1. Circle with a diameter of 18 cm
2. Circle with a radius of 31.5 m
3. Circle with a diameter of 27 mm
4. Circle with a radius of 22.5 m
5. Circle with a diameter of 81 cm
6. Circle with a radius of 54 m
7. Circle with a diameter of 99 mm
8. Circle with a radius of 18 m
9. Circle with a diameter of 72 cm
10. Circle with a radius of 27 m

LESSON 11: AREA OF A RECTANGLE

Introduction

Problem 49 of the RMP (and for that matter Problem 6 of the Moscow Mathematical Papyrus) concerns the area of a rectangle. Both papyri imply a formula of A = h x b where A is the area, h is the height of the rectangle, and b is the base of the rectangle.

Questions

Calculate the areas of these rectangles without a calculator
  1. 12 x 7
  2. 7 x 9
  3. 21 x 4
  4. 9 x 6
  5. 16 x 5
  6. 7 x 7
  7. 8 x 6
  8. 9 x 12
  9. 13 x 14
  10. 16 x 18
  11. The length of one side of a rectangle is 8 cm, the area of the rectangle is 104 cm^2, what is the length of the other side of the rectangle?
  12. The area of a rectangle is 136 m^2, it is 8 m long, what is its height?
  13. The area of a rectangle is 126 cm^2, its height is 12 cm, how long is it?
  14. The length of a rectangle is 12.6 mm, its area is 151.2 mm^2, what is its height?
  15. The area of a rectangle is 288 cm^2, work out 3 possible dimensions for its length and height.
  16. The perimeter of a square is equal to its area, what is the length of this square?
  17. The length of a rectangle is 14 cm, its perimeter is 50 cm, work out its area.
  18. The perimeter of a rectangle is half as much as its area, how long could it be?
  19. The area of a rectangle is 168 m^2. It is 12 m in height. Work out the length.
  20. The area of a rectangle is 96 mm^2. It is 6.4 mm long, what is its height?

LESSON 12: AREA OF A TRIANGLE

Introduction

Problem 51 of the RMP concerns the area of a triangle. The question reads as follows: "A demonstration of the calculation of a triangular plane. If asked: A triangle 10 rods high, 4 at its base; what's its area?" Both papyri, RMP and MMP, imply a formula of A = 1/2 x b x h where A is the area, h is the height of the triangle and b is the base of the triangle.

The solution is given as
4/2 = 2   This is the base halved (or 1/2 x b)
2 x 10 = 20   Base halved multiplied by the height (or 1/2 b x h) gives the area

Questions

Calculate the area of these triangles. The height is given first and then the base.

  1. 12 x 6
  2. 7 x 10
  3. 21 x 5
  4. 9 x 7
  5. 16 x 6
  6. 7 x 8
  7. 8 x 5
  8. 9 x 11
  9. 13 x 13
  10. 16 x 17
  11. The base of a triangle is 7 cm, the area of the triangle is 103 cm², what is its height?
  12. The area of a triangle is 135 m², the base is 8 m long, what is its height?
  13. The area of a triangle is 124 cm², its height is 12 cm, how long is its base?
  14. The base of the triangle is 12.6 mm, its area is 148.2 mm², what is its height?
  15. The area of a triangle is 288 cm², work out 3 possible dimensions for its base and height.

LESSON 13: MATHEMATICAL SERIES

Introduction

The RMP had a question: "Suppose that on an estate of 7 houses, each house had 7 cats, each cat killed 7 mice, each mouse ate 7 barley seeds, and each barley seed would have yielded 7 bushels; how many bushels would that make all told?" The answer was calculated as $7 \times 7 \times 7 \times 7 \times 7 = 16807$.

Questions

1. (a) Calculate the total number of bushels if there were 8 houses, each with 8 cats, each chasing 8 mice, each eating 8 barley seeds, yielding 8 bushels. (b) Convert this number to 8 raised to the correct power.

2. (a) Calculate the total number of bushels if there were 6 houses, each with 6 cats, each chasing 6 mice, each eating 6 barley seeds, yielding 6 bushels. (b) Convert this number to 6 raised to the correct power.

3. (a) Calculate the total number of barley seeds if there were 9 houses, each with 9 cats, each chasing 9 mice, each eating 9 barley seeds. (b) Convert this number to 9 raised to the correct power.

4. (a) Calculate the total number of mice if there were 11 houses, each with 11 cats, chasing 11 mice. (b) Convert this number to 11 raised to the correct power.

5. (a) Calculate the total number of cats if there were 13 houses each with 13 cats. (b) Convert this number to 13 raised to the correct power.

6. (a) Fibonacci wrote a mathematics problem that asks: "Seven old women went to Rome: each woman had seven mules; each mule carried seven sacks: each sack contained seven loaves; and with each loaf were seven knives; each knife was put up in seven sheaths." Calculate the total number of sheathes carried by the old women. (b) Convert this number to 7 raised to the correct power.

7. (a) The Mother Goose rhyme says: "As I was going to St. Ives, I met a man with seven wives. Each wife had seven sacks, each sack had seven cats, each cat had seven kits. Kits, cats, sacks, wives, how many were going to St. Ives?" Calculate the total number of sacks carried by the wives. (b) Convert this number to 7 raised to the correct power.

8. (a) Using the Mother Goose rhyme, calculate the total number of cats. (b) Convert this number to 7 raised to the correct power.

9. (a) Using the Mother Goose rhyme, calculate the total number of kits. (b) Convert this number to 7 raised to the correct power.

10. (a) Using the Mother Goose rhyme, calculate how many people were ACTUALLY GOING to St Ives. (b) Convert this number to 7 raised to the correct power.

LESSON 14: 'PYTHAGOREAN' THEOREM

Introduction

Plutarch, a great Greco-Roman scholar, wrote: "The Egyptians appeared to have figured out the world in the form of the most beautiful of triangles ... This triangle, the most beautiful of triangles, has its vertical side composed of three, its base of four, and its hypotenuse of five parts, and the square of the latter is equal to the sum of the squares of the two sides."

Questions

Using the formulae $a^2 + b^2 = c^2$, $a^2 = c^2 - b^2$ and $b^2 = c^2 - a^2$, find the missing numbers.

1. Side a of a triangle has a length of 6 cm, side b has a length of 8 cm, using the formulae above, what is the length of side c?

2. Side a of a triangle has a length of 5 m, side b has a length of 12 m, using the formulae above, what is the length of side c?

3. Side a of a triangle has a length of 9 mm, side b has a length of 12 mm, using the formulae above, what is the length of side c?

4. Side a of a triangle has a length of 8 cm, side b has a length of 15 cm, using the formulae above, what is the length of side c?

5. Side a of a triangle has a length of 12 m, side b has a length of 16 m, using the formulae above, what is the length of side c?

6. Side a of a triangle has a length of 30 cm, side c has a length of 50 cm, using the formulae above, what is the length of side b?

7. Side b of a triangle has a length of 80 cm, side c has a length of 100 cm, using the formulae above, what is the length of side a?

8. Using graph paper, draw and label the triangles using the measurements for questions 1 and 4.

LESSON 15: THE MATHEMATICS OF AL-HASSAR - THE FRACTION BAR

Introduction

Al-Hassar wrote a book called *The Book of Proof and Recall.* The book is of historical importance because fractions are written symbolically with the horizontal bar and the Hindu numerals--i.e. the ancestors of the digits that we use today. In the chapter on fractions he wrote: "but if you have to represent a fraction, then write the denominator under a [horizontal] line and above each its mentioned part."

Questions

Write the following numbers as fractions. All of your answers must be in the form of a numerator, a fraction bar, and a denominator ONLY.

1. One half
2. A quarter
3. Two thirds
4. One and a third
5. Three and a fifth
6. Six and two thirds
7. Eight and one sixth
8. Eleven and two sevenths
9. Fourteen and three fifths
10. Nineteen and a half
11. Twenty two and three quarters
12. Twenty six and two thirds
13. Twenty nine and a third
14. Thirty three and a fifth
15. Thirty seven and two thirds
16. Thirty eight and five sixths
17. Forty one and three sevenths
18. Forty four and four fifths
19. Forty nine and a half
20. Fifty seven and a quarter

LESSON 16: THE MATHEMATICS OF AL-HASSAR: PRIME NUMBERS

Introduction

Al-Hassar's *The Complete Book on the Art of Number* contains a number of mathematical topics. One issue he dealt with was the decomposition of a number into prime factors.

Questions

Find the prime numbers which add up to these numbers. (You cannot double a prime number to find your answer).

1. 16
2. 64
3. 28
4. 40
5. 24
6. 76
7. 112
8. 56
9. 72
10. 400
11. 200
12. 158
13. 194
14. 206
15. 384
16. 140
17. 134
18. 144
19. 176
20. 514
21. Show 3 examples where a pair of prime numbers adds up to a square number.
22. Show 3 examples where a pair of prime numbers adds up to a cube number.

LESSON 17: THE MATHEMATICS OF AL-MAGHRIBI: POLYHEDRA

Introduction

Muhyi l'din al-Maghribi (1220-1283) was an eminent Moorish astronomer. He wrote a famous commentary on ideas that date back to Plato that covers "the ratios of (1) the edges, (2) the faces, (3) the surface areas, (4) the perpendicular distances from the centre to a face and (5) the volumes of the five regular polyhedra inscribed in one sphere." Polyhedra are solid figures with many plane faces, typically more than six.

A vertex is a point where two or more straight lines meet (plural vertices). A vertex (or V) is a corner. An edge (or E) is a line segment that joins 2 corners (vertices). A face (or F) is one of the individual surfaces of a solid object. There is a formula that connects these three concepts where F + V - E = 2

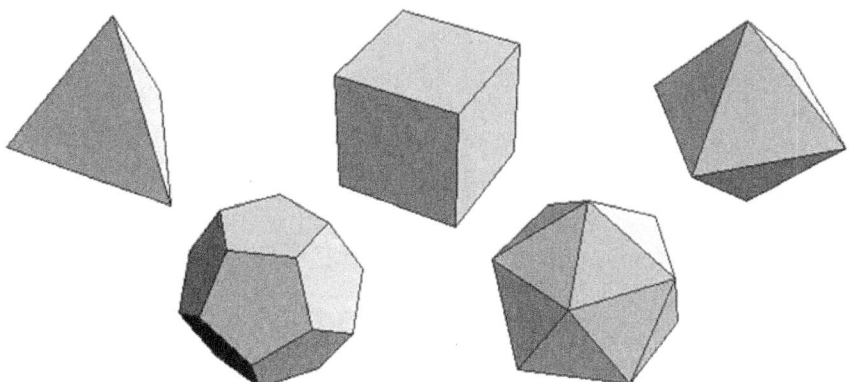

From left to right, the polyhedra are tetrahedron, dodecahedron, cube, icosahedron and octahedron.

Questions

1. (a) How many faces does a cube have? (b) How many vertices (corners) does it have? (c) How many edges does it have? (d) Show that F + V - E = 2 for a cube.

2. A tetrahedron has 4 faces. (a) How many vertices does it have? (b) How many edges does it have? (c) Show that F + V - E = 2 for a tetrahedron.

3. An octahedron has 8 faces. (a) How many vertices does it have? (b) How many edges does it have? (c) Show that F + V - E = 2 for a octahedron.

4. A dodecahedron has 30 edges and 20 vertices. Use the formula to calculate how many faces it has.

5. An icosahedron has 20 faces and 12 vertices. Use the formula to calculate how many edges it has.

6. Use plasticine to model a tetrahedron, an octahedron and a dodecahedron.

LESSON 18: ANCIENT ETHIOPIAN MULTIPLICATION

Introduction

If an Ethiopian trader wanted to multiply 11 by 15, he would put the numbers into two columns. He would place the 11 in one column and he would place the 15 in the other column.

In the first column he would continually halve the number ignoring the fractions. Thus 11 halved is 5 (i.e. ignoring the fractions), halved again is 2, and halved again is 1. In the other column he would double the numbers. Thus 15 doubled is 30, doubled again is 60, and doubled again is 120. The two columns might look like this:

Halving column	Doubling column
11	15
5	30
2	60
1	120

There is a rule that one must IGNORE any even number(s) in the halving column AND the corresponding number(s) in the doubling column. Consequently, we shall ignore the 2 and the 60. Our table now looks like this.

Halving column	Doubling column
11	15
5	30
1	120

Finally we add up the numbers in the doubling column to produce our answer which is 15 + 30 + 120 = 165.

Questions

Use this method to multiply the following numbers together. Put the first number in the halving column and put the second number in the doubling column.

1. 3 x 5
2. 6 x 8
3. 9 x 14
4. 12 x 16
5. 17 x 23
6. 19 x 38
7. 24 x 54
8. 26 x 61
9. 37 x 101
10. 3 x 6.5
11. 6 x 9.4
12. 9 x 15.3
13. 12 x 17.2
14. 17 x 23.1
15. 20 x 39.44
16. 25 x 55.63
17. 27 x 62.23
18. 38 x 102.74
19. 41 x 20.34
20. 20.34 x 41

Discussion Questions

1. Did you find this method faster or slower than the Ancient Egyptian method? Give a reason for your answer.

2. Why are the answers to questions 19 and 20 slightly different to each other? Which one is the correct answer?

Note

For this method to produce the correct answer, the number in the halving column MUST BE A WHOLE NUMBER!

LESSON 19: TCHOKWE NETWORKS AND THE GREATEST COMMON DIVISOR

Introduction

A traditional sand design of the Tchokwe of Angola was to trace an antelope without taking your pen or pencil off the paper and without retracing the same line. Using squared paper, draw 12 points consisting of 3 rows of 4 points one centimetre apart. Trace the antelope design. Do not take your pen or pencil off the paper and you should not retrace the same line. Finally you can add the head, the tail and the legs ... if you wish.

Questions

Draw a series of antelope designs without the head, tail and legs and complete the following table.
  1. 2 rows by 2 columns
  2. 2 rows by 3 columns
  3. 3 rows by 3 columns
  4. 3 rows by 4 columns
  5. 4 rows by 2 columns
  6. 4 rows by 3 columns
  7. 4 rows by 4 columns
  8. 4 rows by 5 columns
  9. 5 rows by 3 columns
  10. 5 rows by 6 columns

Rows	Columns	How many complete lines?
2	2	2
2	3	
3	3	
Etcetera		

Note

The number of complete lines is the greatest common divisor of the number of rows and the columns.

Questions

11. What is the greatest common divisor of 6 and 10? Prove your answer by drawing the correct antelope design.

12. What is the greatest common divisor of 10 and 12? Prove your answer by drawing the correct antelope design.

13. What is the greatest common divisor of 12 and 16? Prove your answer by drawing the correct antelope design.

14. What is the greatest common divisor of 15 and 20? Prove your answer by drawing the correct antelope design.

15. What is the greatest common divisor of 17 and 20? Prove your answer by drawing the correct antelope design.

LESSON 20: MORE ON TCHOKWE NETWORKS

Introduction

The Tchokwe antelope design has 3 rows of 4 points and indicates that 3 x 4 = 12. The points can also be counted along the diagonals from left to right as 1, 2, 3, 3, 2 and 1. This indicates that 3 x 4 = 12 = (1 + 2 + 3) + (3 + 2 + 1). The second side of the equation can be rearranged as (1 + 2 + 3) + (1 + 2 + 3) or even as 2 x (1 + 2 + 3).

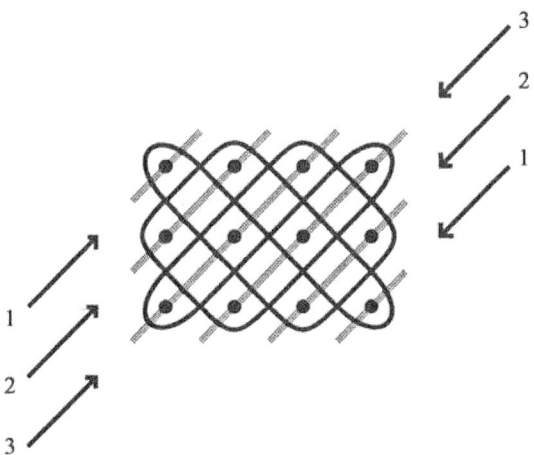

Since:
2 x (1 + 2 + 3) = 3 x 4 = 12

Therefore:
1 + 2 + 3 = (3 x 4)/2 = 6

Questions

1. Draw an antelope design of 5 by 6 points. Annotate on the design like the example above to show that 5 x 6 = (1 + 2 + 3 + 4 + 5) + (5 + 4 + 3 + 2 + 1).

2. Draw an antelope design of 8 by 9 points. Annotate on the design to show that 8 x 9 = (1 + 2 + 3 + 4 + 5 + 6 + 7 + 8) + (8 + 7 + 6 + 5 + 4 + 3 + 2 + 1).

Note

If 5 x 6 = (1 + 2 + 3 + 4 + 5) + (5 + 4 + 3 + 2 + 1) then 5 x 6 = 2 x (1 + 2 + 3 + 4 + 5).

   Also 1 + 2 + 3 + 4 + 5 = (5 x 6)/2 = 15

Questions

   3. What is 1 + 2 + 3 + 4 + 5 + 6 + 7 + 8?

   4. Suggest two ways of how you could have calculated this.

Note

The antelope designs suggest that: n x (n + 1) = 2 x (1 + 2 + 3 + ... + n) also n x (n + 1)/2 = 1 + 2 + 3 + ... + n.

   Suppose a mathematics problem required you to add all numbers from 1 to 100. We could laboriously add up 1 + 2 + 3 + ... + 99 + 100 or we could use the short cut n x (n + 1)/2 or (100 x 101)/2 = 5050.

Questions

   5. Add up all natural numbers from 1 to 65.

   6. Add up all natural numbers from 1 to 120.

   7. Add up all natural numbers from 1 to 189.

   8. Add up all natural numbers from 1 to 215.

   9. Add up all natural numbers from 1 to 268.

   10. Add up all natural numbers from 1 to 651.

   11. Add up all natural numbers from 66 to 99.

   12. Add up all natural numbers from 100 to 120.

   13. Add up all natural numbers from 43 to 111.

   14. Add up all natural numbers from 112 to 650.

   15. Add up all natural numbers from 1 to 999.

LESSON 21: MAGIC SQUARES

Introduction

How is it possible to arrange numbers into a table using each number only once so that each row, each column, and the two diagonals all add up to the same number? This question had intrigued Chinese scholars for thousands of years. However, West African scholars were also interested in this puzzle. Not only did they produce solutions to this problem, they even wrote examples as good luck charms and benedictions!

n+3	n-4	n+1
n-2	n	n+2
n-1	n+4	n-3

Tasks

Using the above formula, generate 16 magic squares where

1. (a) $n = 5$ (b) reflect this square about the vertical axis (c) calculate the magic constant

2. (a) $n = 6$ (b) rotate this square by 90 degrees clockwise (c) calculate the magic constant

3. (a) $n = 8$ (b) reflect this square about the horizontal axis (c) calculate the magic constant

4. (a) $n = 11$ (b) rotate this square by 180 degrees (c) calculate the magic constant

5. (a) $n = 14$ (b) reflect this square about the 45 degree diagonal axis (c) calculate the magic constant

6. (a) $n = 19$ (b) rotate this square by 270 degrees clockwise (c) calculate the magic constant

7. (a) $n = 25$ (b) rotate this square about the 135 degree diagonal axis (c) calculate the magic constant

8. (a) $n = 31$ (b) rotate this square by 90 degrees anticlockwise then reflect it about the vertical axis (c) calculate the magic constant

ANSWERS

Lesson 1: Ibrahima and Goldbach's Conjecture

1. 3 and 1
2. 7 and 5
3. 13 and 1, 11 and 3
4. 11 and 5, 13 and 3
5. 17 and 1, 13 and 5, 11 and 7
6. 17 and 3, 13 and 7
7. 17 and 5, 11 and 11
8. 23 and 1, 19 and 5, 13 and 11
9. 23 and 3, 19 and 7, 13 and 13
10. 23 and 5, 17 and 11
11. 31 and 1, 29 and 3, 19 and 13
12. 31 and 3, 29 and 5, 23 and 11
13. 37 and 1, 31 and 7
14. 37 and 5, 31 and 11, 29 and 13, 23 and 19,
15. 43 and 3, 41 and 5, 29 and 17, 23 and 23
16. 47 and 3, 43 and 7, 37 and 13, 31 and 19
17. 53 and 1, 47 and 7, 43 and 11, 41 and 13, 31 and 23
18. 53 and 5, 47 and 11, 29 and 29
19. 61 and 1, 59 and 3, 43 and 19, 31 and 31
20. 61 and 5, 59 and 7, 53 and 13, 47 and 19, 43 and 23, 37 and 29
21. 67 and 3, 59 and 11, 53 and 17, 47 and 23, 41 and 29
22. 73 and 1, 71 and 3, 67 and 7, 61 and 13, 43 and 31, 37 and 37
23. 73 and 5, 71 and 7, 67 and 11, 61 and 17, 59 and 19, 47 and 31
24. 79 and 3, 71 and 11, 59 and 23, 53 and 29, 41 and 41
25. 83 and 3, 79 and 7, 73 and 13, 67 and 19, 43 and 43
26. 89 and 1, 83 and 7, 79 and 11, 73 and 17, 71 and 19, 67 and 23, 61 and 29, 59 and 31, 53 and 37, 47 and 43
27. 89 and 5, 83 and 11, 71 and 23, 53 and 41, 47 and 47
28. 97 and 1, 79 and 19, 71 and 17, 67 and 31, 61 and 37
29. 101 and 1, 97 and 5, 89 and 13, 83 and 19, 79 and 23, 73 and 29, 71 and 31, 61 and 41, 59 and 43
30. 103 and 3, 101 and 5, 89 and 17, 83 and 23, 59 and 47, 53 and 53

Lesson 2: The Lebombo Bone

1. 89	2. 236	3. 207	4. 148	5. 531
6. 797	7. 1416	8. 1888	9. 944	10. 2360
11. 24	12. 14	13. 18	14. 40	15. 266
16. 384	17. 472	18. 2124	19. 144	20. 52

Lesson 3: The Ishango Bone

1.
4 + 8 = 12, which is 16 - 4
8 + 8 = 16
17 + 3 = 20, which is 16 + 4
5 + 11 = 16

2.
7 + 7 = 14, which is 21 - 7
10 + 11 = 21
21 + 7 = 28, which is 21 + 7
9 + 12 = 21

3.
8 + 16 = 24, which is 36 - 12
12 + 24 = 36
40 + 8 = 48, which is 36 + 12
5 + 31 = 36

4.
1 + 4 = 5, which is 10 - 5
6 + 4 = 10
10 + 5 = 15, which is 10 + 5
5 + 5 = 10

5.
2 + 8 = 10, which is 20 - 10
13 + 7 = 20
20 + 10 = 30, which is 20 + 10
8 + 12 = 20

6.
9 + 16 = 25, which is 50 - 25
20 + 30 = 50
39 + 36 = 75, which is 50 + 25
5 + 45 = 50

7.
6 + 29 = 35, which is 42 - 7
28 + 14 = 42
42 + 7 = 49, which is 42 + 7
30 + 12 = 42

8.
17 + 39 = 56, which is 64 - 8
24 + 40 = 64
49 + 23 = 72, which is 64 + 8
5 + 59 = 64

9.
30 + 60 = 90, which is 120 - 30
40 + 80 = 120
100 + 50 = 150, which is 120 + 30
108 + 12 = 120

10.
23 + 85 = 108, which is 144 - 36
60 + 84 = 144
97 + 73 = 170, which is 144 + 36
5 + 139 = 144

Lesson 4: Egyptian numerals

1. 140331	2. 102303
3. 131036	4. 1011102
5. 116059	6. 101130
7. 5300020	8. 40100
9. 1031000	10. 2010000

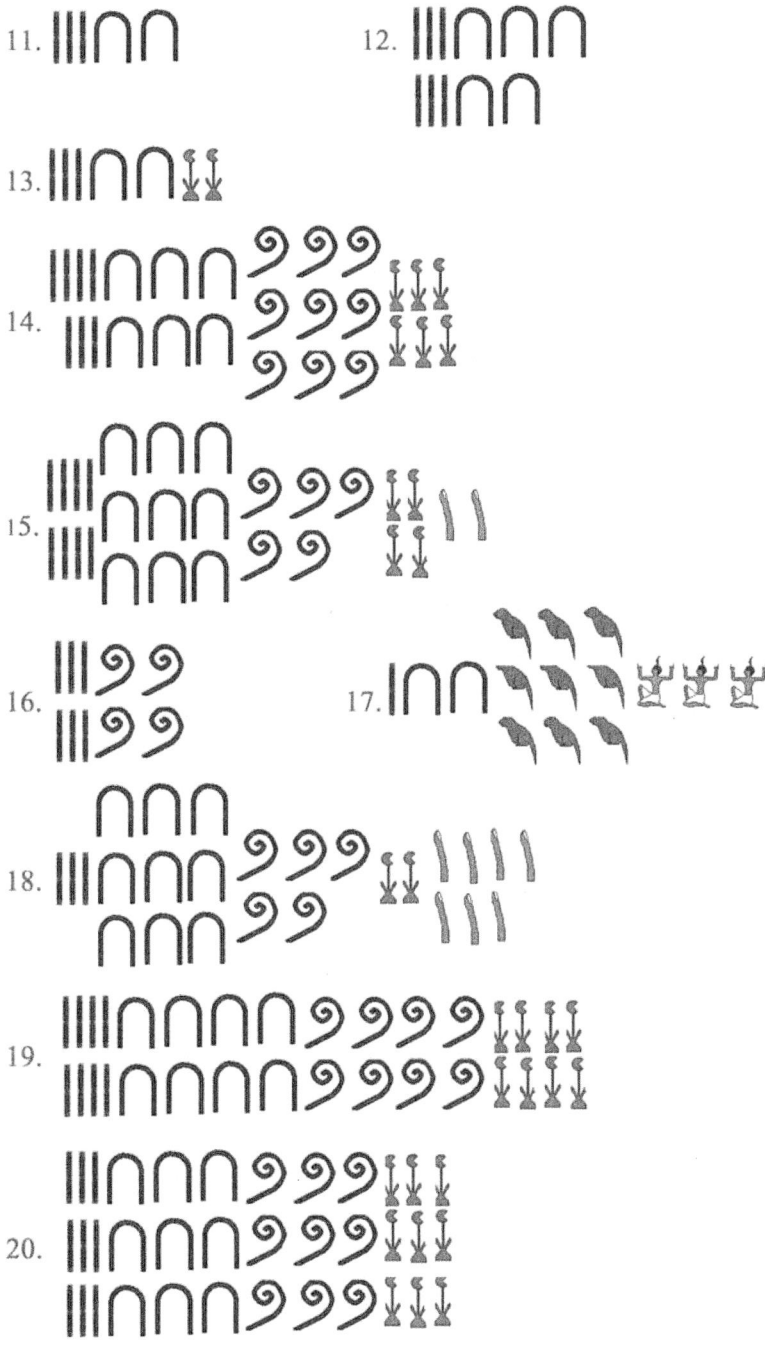

Lesson 5: Unit Fractions

1. (a) 7/8 (b) 0.875   2. (a) 7/16 (b) 0.4375
3. (a) 1/2 (b) 0.5   4. (a) 7/12 (b) 0.58333
5. (a) 5/8 (b) 0.625   6. (a) 11/32 (b) 0.34375
7. (a) 63/64 (b) 0.984375   8. (a) 7/20 (b) 0.35
9. (a) 1/3 (b) 0.333   10. (a) 13/45 (b) 0.2888
11. (a) 31/80 (b) 0.3875   12. (a) 1/4 (b) 0.25
13. (a) 11/12 (b) 0.91666   14. (a) 11/24 (b) 0.458333
15. (a) 17/30 (b) 0.5666   16. (a) 17/60 (b) 0.28333
17. (a) 13/35 (b) 0.3714285   18. (a) 23/70 (b) 0.3285714
19. (a) 13/63 (b) 0.206349   20. (a) 76/126 = 38/63 (b) 0.603174

Lesson 6: Ancient Egyptian Multiplication

1. 12   2. 24   3. 117   4. 208   5. 527
6. 684   7. 1564   8. 7003   9. 7751.8
10. Cannot be done

1. 12   2. 24   3. 117   4. 208   5. 527
6. 684   7. 1564   8. 7003   9. Cannot be done
10. 102602.5

Lesson 7: Ancient Egyptian Division

1. 3   2. 4
3. 9   4. 9
5. 8   6. 10 remainder 3
7. 19 remainder 2   8. 22
9. 21 remainder 12   10. 22 remainder 7

Lesson 8: The Cylinder

1. 2560 mm^3 2. 62720 cm^3
3. 2880 m^3 4. 64000 mm^3
5. 414720 m^3 6. 1843200 cm^3
7. 1393920 m^3 8. 51200 mm^3
9. 368640 m^3 10. 138240 cm^3

1. 2544.6879 m³   2. 62344.85355 cm³
3. 2862.773888 m³   4. 63617.1975 mm³
5. 412239.4398 m³   6. 1832175.288 cm³
7. 1385582.562 m³   8. 50893.758 mm³
9. 366435.0576 m³   10. 137413.1466 cm³

Lesson 9: Cubing a number

1. 1   2. 8
3. 64   4. 125
5. 512   6. 1331
7. 2744   8. 4913
9. 9261   10. 17576
11. 29791   12. 50653
13. 85184   14. 140608
15. 226981   16. 357911
17. 551368   18. 830584
19. 1225043   20. 1771561

Lesson 10: Area of a Circle

1. 256 cm²   2. 3136 m²
3. 576 mm²   4. 1600 m²
5. 5184 cm²   6. 9216 m²
7. 7744 mm²   8. 1024 m²
9. 4096 cm²   10. 2304 m²

1. 254.46879 cm²   2. 3117.242678 m²
3. 572.5547775 mm²   4. 1590.429938 m²
5. 5152.992998 cm²   6. 9160.87644 m²
7. 7697.680898 mm²   8. 1017.87516 m²
9. 4071.50064 cm²   10. 2290.21911 m²

Lesson 11: Area of a Rectangle

1. 84
2. 63
3. 84
4. 54
5. 80

6. 49
7. 48
8. 108
9. 182
10. 288
11. 13 cm
12. 17 m
13. 10.5 cm
14. 12 mm
15. Any three pairs of numbers that multiply to 288 cm^2
16. Each side is 4 units in length. Thus the area is 4 x 4
17. 154 cm^2.
18. 20 x 5. The area is 100 units. The perimeter is 50 units.
19. 14 m
20. 15 mm

Lesson 12: Area of a Triangle

1. 36	2. 35
3. 52.5	4.31.5
5. 48	6. 28
7. 20	8. 49.5
9. 84.5	10. 136
11. 29.42 cm	12. 33.75 m
13. 20.66 cm	14. 23.52 mm

15. Any possible pairs of numbers that multiply to 576 cm^2

Lesson 13: Mathematical Series

1. (a) 32768 bushels (b) 8 to the power of 5
2. (a) 7776 bushels (b) 6 to the power of 5
3. (a) 6561 barley seeds (b) 9 to the power of 4
4. (a) 1331 mice (b) 11 to the power of 3
5. (a) 169 cats (b) 13 to the power of 2
6. (a) 117649 sheathes (b) 7 to the power of 6
7. (a) 49 sacks (b) 7 to the power of 2
8. (a) 343 cats (b) 7 to the power of 3
9. (a) 2401 kits (b) 7 to the power of 4
10. (a) 1 (b) 7 to the power of 0

Lesson 14: Pythagorean Theorem

1. 10 cm	2. 13 cm
3. 15 mm	4. 17 cm
5. 20 m	6. 40 cm
7. 60 cm	

Lesson 15: The Mathematics of Al-Hassar: The Fraction Bar

1. 1/2	2. 1/4	3. 2/3	4. 4/3	5. 16/5
6. 20/3	7. 49/6	8. 79/7	9. 73/5	10. 39/2
11. 91/4	12. 80/3	13 88/3	14. 166/5	15 113/3
16. 233/6	17. 290/7	18. 224/5	19. 99/2	20. 229/4

Lesson 16: The Mathematics of Al-Hassar: Prime Numbers

1. Some examples are 11 and 5, 13 and 3
2. 61 and 3, 23 and 41
3. 23 and 5, 11 and 17
4. 23 and 17, 37 and 3
5. 17 and 7, 13 and 11
6. 73 and 3, 53 and 23
7. 89 and 23, 59 and 53
8. 157 and 43, 127 and 73
9. 97 and 61, 139 and 19
10. 181 and 13, 191 and 3
11. 199 and 7, 139 and 67
12. 353 and 31, 191 and 193
13. 67 and 73, 103 and 37
14. 97 and 37, 103 and 31
15. 373 and 11, 191 and 193
16. 73 and 67, 103 and 37
17. 127 and 7, 131 and 3
18. 127 and 17, 107 and 37
19. 139 and 37, 173 and 3
20. 503 and 11
21. Some examples of these are 29 + 7 = 36, 47 + 97 = 144, 59 + 41 = 100, 137 + 59 = 196, 251 + 5 = 256
22. Some examples of these are 3 + 5 = 8, 47 + 17 = 64, 197 + 19 = 216, 499 + 13 = 512, 443 + 557 = 1000

Lesson 17: The Mathematics of al-Maghribi: Polyhedra

1. (a) 6 (b) 8 (c) 12 (d) 6 + 8 - 12 = 2
2. (a) 4 (b) 6 (c) 4 + 4 - 6 = 2
3. (a) 6 (b) 12 (c) 8 + 6 - 12 = 2
4. 12
5. 30

Lesson 18: Ancient Ethiopian Multiplication

1. 15	2. 48
3. 126	4. 192
5. 391	6. 722
7. 1296	8. 1586
9. 3737	10. 19.5
11. 56.4	12. 137.7
13. 206.4	14. 392.7
15. 788.8	16. 1390.75
17. 1680.21	18. 3904.12

19. 833.94
20. 820 However, the Ethiopian method produced this incorrect answer

Lesson 19: Tchokwe Networks and the Greatest Common Divisor

The number of complete lines (or the greatest common divisor) are given below:

1. 2	2. 1	3. 3	4. 1	5. 2
6. 1	7. 4	8. 1	9. 1	10. 1
11. 2	12. 2	13. 4	14. 5	15. 1

Lesson 20: More on Tchokwe Networks

3. 36	4. (i) Add the numbers up or (ii) use the formula
5. 2145	6. 7260
7. 17955	8. 23220
9. 36046	10. 212226
11. 4950 - 2145 = 2805	12. 7260 - 4950 = 2310
13. 6216 - 903 = 5315	14. 423150 - 6216 = 417024
15. 499500	

Lesson 21: Magic Squares

1. (a)

8	1	6
3	5	7
4	9	2

(b)

6	1	8
7	5	3
2	9	4

(c) The magic constant is 15

2. (a)

9	2	7
4	6	8
5	10	3

(b)

5	4	9
10	6	2
3	8	7

(c) The magic constant is 18

3. (a)

11	4	9
6	8	10
7	12	5

(b)

7	12	5
6	8	10
11	4	9

(c) The magic constant is 24

4. (a)

14	7	12
9	11	13
10	15	8

(b)

8	15	10
13	11	9
12	7	14

(c) The magic constant is 33

5. (a)

17	10	15
12	14	16
13	18	11

(b)

11	16	15
18	14	10
13	12	17

(c) The magic constant is 42

6. (a)

22	15	20
17	19	21
18	23	16

(b)

20	21	16
15	19	23
22	17	18

(c) The magic constant is 57

7. (a)

28	21	26
23	25	27
24	29	22

(b)

28	23	24
21	25	29
26	27	22

(c) The magic constant is 75

8. (a)

34	27	32
29	31	33
30	35	28

(b)

28	33	32
35	31	27
30	29	34

(c) The magic constant is 93

PART THREE

THINK LIKE A GENIUS

INTRODUCTION

'Mathematics is the queen of the sciences and arithmetic is the queen of mathematics'
- Carl Friedrich Gauss

Too many learners experience consternation when faced with mental calculations. In addition, many neglect to practice this essential skill since they assume that the ability to be proficient with numbers is unattainable to them. I have witnessed a positive change in people being pleasantly surprised by their own latent abilities after they have been taught new methods for calculating number problems.

Contrary to popular misconceptions, you don't have to be a 'mathematics genius' or somebody who 'was just born good at mathematics' to be able to become proficient in mental arithmetic. Neuro linguistic programming advocates the notion that any skill can be learned by those who desire to do so. This is certainly true with mathematics.

I have seen people who proclaim to hate mathematics suddenly change when their hidden genius suddenly awakes after sleeping since their childhood. Learners have an irrational fear of number problems which I believe is cultivated in childhood due to the limited methods used to teach mathematics in the education system.

However, there are a plethora of techniques which could be used by learners to simplify the process of seemingly complex calculations. The learner who knows how to employ a number and variety of techniques and strategies can approach mental arithmetic, as well as other tasks, with a confidence that those who rely on only one method do not possess.

When a parent teaches a child to read, the child will have to learn about the sound that each letter makes before they are able to read and comprehend full sentences.

Similarly, when learning the piano, learners are not taught to play pieces first, they are taught the basics of notes and keys. Once a piano student has acquired a good knowledge of the keys of the piano, they can use this knowledge to play music and even compose their own. Learning about an individual number is like learning to play the basic notes on a piano.

Using this same principle, I will show the unique character that some of the individual numbers possess as well as their relationships with other numbers.

To begin the process of re-awakening your hidden genius, you must remember these three key principles:

1. Numbers are like onions. They have many layers. The factors of a number are its layers. When calculating a sum, an understanding of its factors (with the exception of prime numbers which can only be divided by 1 and itself) will assist you in calculating the answer.

2. All big things are made of small things. This is also true of 'large numbers'.

3. Place value is powerful. Place value can change a large number into a small number.

Imagine that you are in possession of the world's most advanced computer. This computer has the potential to do amazing things. It has an amazing memory and could help you to unlock abilities in you which you thought you never had.

Imagine how you would feel if you had this computer but were not taught how to use it properly. Sadly, for many of us, this scenario is the reality.

We do indeed possess the world's most powerful computer but we don't exercise it enough to even a fraction of its true potential. The human mind is a mind-blowing (pardon the pun) piece of hardware. You can programme this hardware and make it do things which you never thought were possible. Mental calculations are not exceptions to this rule.

What follows in this section is a sample of the techniques that I use on this subject. For more information feel free to contact me at john_matthrews60@hotmail.co.uk

John Matthews

LESSON 1: THE MAGIC OF 9

The number 9 is unique.

1 x 9 = 9	2 x 9 = 18
3 x 9 = 27	4 x 9 = 36
5 x 9 = 45	6 x 9 = 54
7 x 9 = 63	8 x 9 = 72
9 x 9 = 81	10 x 9 = 90
11 x 9 = 99	12 x 9 = 108

All multiples of 9 add up to 9.
18 = 1 + 8 = 9
63 = 6 + 3 = 9

If we make 9 our multiplicand an interesting pattern emerges.
*For this exercise all numbers are multiplied by 9

Number multiplied by 9	Sum will end in
1	9
2	8
3	7
4	6
5	5
6	4
7	3
8	2
9	1
10	0

Notice that the number multiplied by 9 and its result always adds up to 10.
The same principle applies when dividing by 9.

LESSON 2: 'BIG NUMBERS' ARE SMALL NUMBERS

To illustrate this principle, I will use the number 72--one of my favourite numbers.

How would I mentally calculate 72 x 7?

I reframe the number 72 to mean '7 tens and 2 units'. Multiplying 7 and 2 is MUCH easier than multiplying 72.

7 x 7 = 49 tens
7 x 2 = 14 units
Thus, 72 x 7 becomes '49 tens and 14 units'. 49 tens added to 14 units gives us 504.

This reframing technique may be used to multiply 3-digit numbers. For example:
702 x 7 becomes '49 hundreds, 0 tens and 14 units'.
This will give the correct answer of 4914.

Here are other examples. 32 x 9 can be seen as 27 tens and 18 units (= 288). Moreover 68 x 7 can be seen as 42 tens and 56 units (= 476). Also 64 x 8 can be seen as 48 tens and 32 units (= 512).

Using this concept, work out the following sums:

1. 46 x 9	2. 58 x 6
3. 53 x 8	4. 36 x 9
5. 37 x 8	6. 63 x 8
7. 48 x 9	8. 56 x 9
9. 72 x 6	10. 76 x 5
11. 34 x 4	12. 95 x 9
13. 96 x 7	14. 88 x 8
15. 33 x 9	16. 22 x 8
17. 81 x 4	18. 44 x 9
19. 68 x 3	20. 84 x 8

LESSON 3: DOUBLING AND HALVING

35 x 18 may appear to be a complex calculation, however, it is possible to 'represent' this sum in a way that is easier to digest. It becomes 35 x 2 = 70 and 18 ÷ 2 = 9
Thus 70 x 9 = 630

Thus by doubling one number and halving the other, a sum emerges which is much easier to handle!

For the sake of clarity, this exercise will involve multiplying by numbers with a unit number of 5.

For example, 45 x 16 becomes 45 x 2 = 90 and 16 ÷ 2 = 8
Thus 90 x 8 = 720

Another example is 15 x 24 becomes 15 x 2 = 30 and 24 ÷ 2 = 12
Thus 30 x 12 = 360

Using this procedure calculate the following
1. 35 x 16	2. 45 x 12
3. 55 x 18	4. 15 x 16
5. 35 x 12	6. 45 x 18
7. 15 x 18	8. 25 x 22
9. 35 x 24	10. 15 x 22
11. 15 x 12	12. 15 x 8
13. 35 x 14	14. 45 x 14
15. 35 x 8	16. 55 x 12
17. 15 x 6	18. 15 x 14
19. 15 x 22	20. 35 x 6

LESSON 4: SUBTRACTING REVERSED NUMBERS

An interesting pattern emerges when reversed numbers are taken away from each other.

Work out the following sums and time yourself doing it. Keep a record of the time because you will need to make a comparison between this exercise and the final exercise.

1. 91 - 19	2. 82 - 28
3. 86 - 68	4. 73 - 37
5. 94 - 49	6. 65 - 56
7. 41 - 14	

Notice that the answers to these sums are always multiples of 9.

The number is always the difference between the 2 numbers multiplied by 9. For instance, the difference between 84 and 48 is $9(8 - 4) = 36$

A = number in the 10s column
B = number in the units column

$AB - BA = 9(A - B)$

Using this process, work out the following sums and time yourself.

1. 96 - 69	2. 85 - 58
3. 76 - 67	4. 92 - 29
5. 62 - 26	6. 21 - 12
7. 93 - 39	

Did your time improve? Can someone explain why the answers are multiples of nine?

ANSWERS

Lesson 2: 'Big numbers' are small numbers

1. 46 x 9 = 414   2. 58 x 6 = 348
3. 53 x 8 = 424   4. 36 x 9 = 324
5. 37 x 8 = 296   6. 63 x 8 = 504
7. 48 x 9 = 432   8. 56 x 9 = 504
9. 72 x 6 = 432   10. 76 x 5 = 380
11. 34 x 4 = 136   12. 95 x 9 = 855
13. 96 x 7 = 672   14. 88 x 8 = 704
15. 33 x 9 = 297   16. 22 x 8 = 176
17. 81 x 4 = 324   18. 44 x 9 = 396
19. 68 x 3 = 204   20. 84 x 8 = 672

Lesson 3: Doubling and halving

1. 35 x 16 = 560   2. 45 x 12 = 540
3. 55 x 18 = 990   4. 15 x 16 = 240
5. 35 x 12 = 420   6. 45 x 18 = 810
7. 15 x 18 = 270   8. 25 x 22 = 550
9. 35 x 24 = 840   10. 15 x 22 = 330
11. 15 x 12 = 180   12. 15 x 8 = 120
13. 35 x 14 = 490   14. 45 x 14 = 630
15. 35 x 8 = 280   16. 55 x 12 = 660
17. 15 x 6 = 90   18. 15 x 14 = 210
19. 15 x 22 = 330   20. 35 x 6 = 210

Lesson 4: Subtracting reversed numbers

1. 96 - 69 = 27   2. 85 - 58 = 27
3. 76 - 76 = 9   4. 92 - 29 = 63
5. 62 - 26 = 36   6. 21 - 12 = 9
7. 93 - 39 = 54

1. 91 - 19 = 72   2. 82 - 28 = 54
3. 86 - 68 = 18   4. 73 - 37 = 36
5. 94 - 49 = 45   6. 65 - 56 = 9
7. 41 - 14 = 27

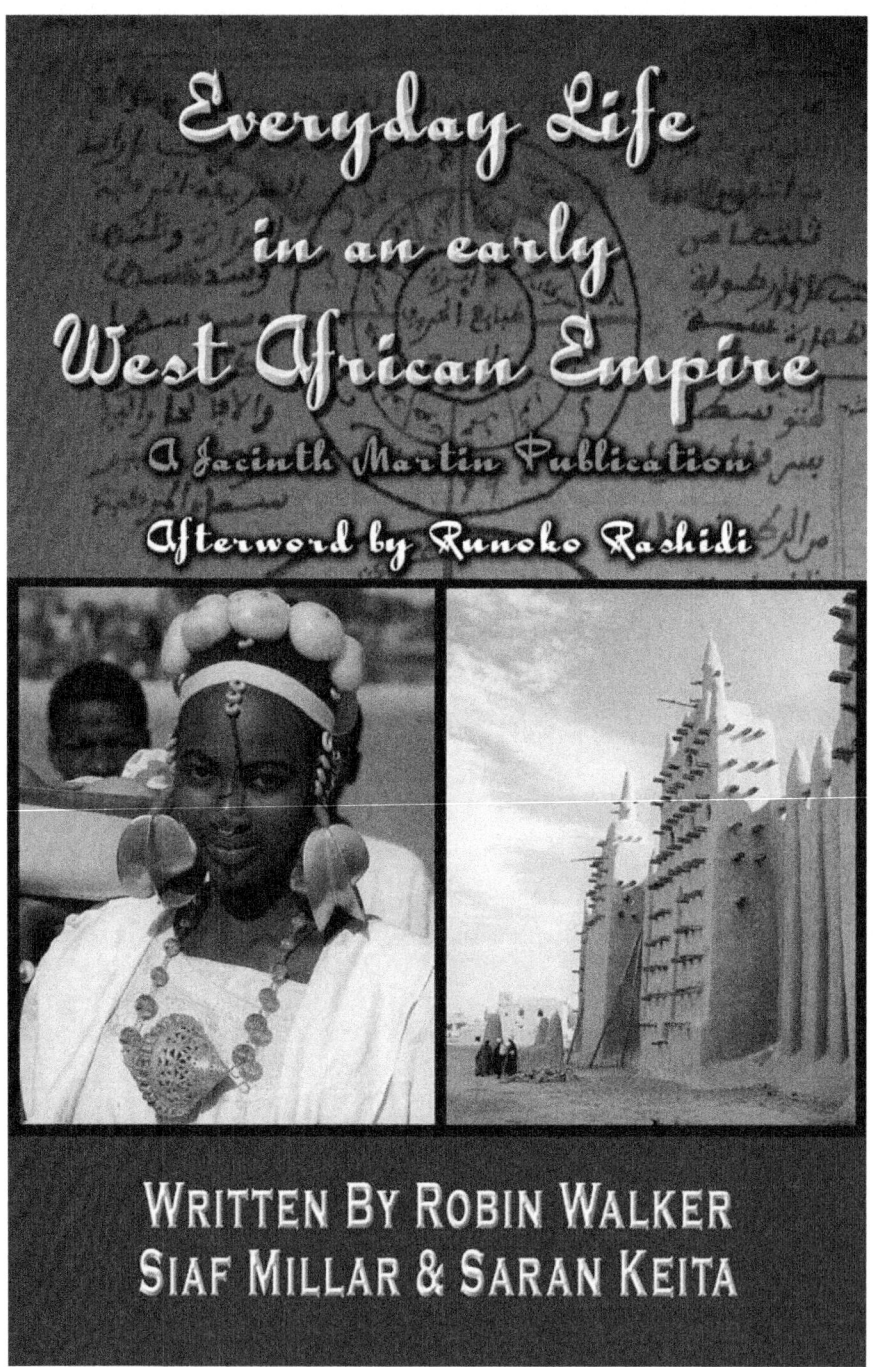

Everyday Life
in an early
West African Empire
A Jacinth Martin Publication

Afterword by Runoko Rashidi

WRITTEN BY ROBIN WALKER
SIAF MILLAR & SARAN KEITA

PART FOUR

AIN'T THAT PECULIAR?

INTRODUCTION

In this section I will share a sample of the number patterns which have baffled and amazed me. Why these patterns exist appears to be a mystery. They also demonstrate the fact that mathematics is truly a unique topic.

Just as the English vocabulary has been expanded incrementally with each passing year, new mathematical concepts are being recognised all the time. As you expand your thinking on the topic, it is my hope that you will recognise a mathematical pattern which will ignite a flame of creative thinking in your own mind.

Considering that we only use a very small percentage of our brain, there yet remains a vast sea of mathematical knowledge which remains uncharted.

As before, this is a small sample of the puzzles that interest me. For more information feel free to contact me at john_matthews60@hotmail.co.uk

John Matthews

LESSON 1: CHANGING UNITS

Take the following examples.

1. 12 x 9 = 108 and 19 x 2 = 38
2. 17 x 9 = 153 and 19 x 7 = 133
3. 14 x 9 = 126 and 19 x 4 = 76

When the units are swapped around in each pair of sums, a very interesting pattern emerges. Do you recognise the pattern?

Here is the formula to calculate the difference between the 2 sums when the units are changed.
T = number in 10s column
A = higher unit
B = lower unit
The formula is 10T x (A - B)

For instance if the sums were (24 x 6) and (26 x 4), then:
6 - 4 = 2
2 x 20 = 40
This means that the difference between (24 x 6) and (26 x 4) is 40.

Consider these examples.

1. 29 x 3 = 87 and 23 x 9 = 207
2. 27 x 4 = 108 and 24 x 7 = 168
3. 36 x 4 = 144 and 34 x 6 = 204
4. 38 x 2 = 76 and 32 x 8 = 256
5. 35 x 7 = 245 and 37 x 5 = 185
6. 48 x 9 = 432 and 49 x 8 = 392
7. 44 x 9 = 396 and 49 x 4 = 196
8. 42 x 8 = 336 and 48 x 2 = 96
9. 72 x 4 = 288 and 74 x 2 = 148
10. 77 x 9 = 693 and 79 x 7 = 553
11. 74 x 6 = 444 and 76 x 4 = 304

The difference between the pair of numbers will always be:
(i) the largest unit subtract the smallest unit
(ii) mutiplied by the number in the 10s column
(iii) mutiplied by 10.

Can somebody explain why or how this pattern occurs? Ain't that peculiar?

Complete the following exercise

Using the formula above, work out the differences between these pairs of numbers

1. 47 x 4 and 44 x 7
2. 12 x 4 and 14 x 2
3. 16 x 3 and 13 x 6
4. 15 x 9 and 19 x 5
5. 27 x 3 and 23 x 7
6. 81 x 5 and 85 x 1
7. 75 x 3 and 73 x 5
8. 45 x 9 and 49 x 5
9. 96 x 3 and 93 x 6
10. 28 x 7 and 27 x 8
11. 12 x 3 and 13 x 2
12. 14 x 7 and 17 x 4
13. 24 x 9 and 29 x 4
14. 19 x 3 and 13 x 9
15. 56 x 9 and 59 x 6
16. 37 x 3 and 33 x 7
17. 36 x 9 and 39 x 6
18. 72 x 8 and 78 x 2
19. 64 x 9 and 69 x 4
20. 22 x 9 and 29 x 2

LESSON 2: YEARS AND MONTHS PART ONE

For the sake of clarity, I have chosen the above title for this section of the book. It could have just as easily been called 'Feet and Inches'.

I suspect that this pattern to which I shall refer occurs because the base number used is 12. Two patterns occur when the given principles are applied.

Take a number of years and months and make the total of years added to the total of months add up to a multiple of 11.

For example, 14 years and 8 months (14 + 8 = 22 which is also 2 x 11)

For each example given, work out the total amount of months.

1. 24 years and 9 months
2. 72 years and 5 months
3. 46 years and 9 months
4. 8 years and 3 months

Notice that the total amount of months in each example is always a multiple of 11.

1. (24 x 12) + 9 = 11 x 27
2. (72 x 12) + 5 = 11 x 77
3. (46 x 12) + 9 = 11 x 51
4. (8 x 12) + 3 = 9 x 11

If the total amount of months is a multiple of 11, the sum of the total of years and months is also a multiple of 11.

Can somebody explain why or how this pattern occurs? Ain't that peculiar?

Complete the following exercises

Convert these amounts of years and months into months.
1. 36 years and 8 months
2. 10 years and 1 month
3. 3 years and 8 months
4. 7 years and 4 months
5. 20 years and 2 months
6. 25 years and 8 months
7. 56 years and 10 months
8. 1 year and 10 months
9. 9 years and 2 months
10. 2 years and 9 months

Convert these amounts of months into years and months.
1. 198 months
2. 253 months
3. 209 months
4. 429 months
5. 44 months
6. 484 months
7. 363 months
8. 121 months
9. 561 months
10. 165 months

LESSON 3: YEARS AND MONTHS PART TWO

To appreciate how this pattern works, you must become familiar with the 13 times table.

1 x 13 = 13	11 x 13 = 143
2 x 13 = 26	12 x 13 = 156
3 x 13 = 39	13 x 13 = 169
4 x 13 = 52	14 x 13 = 182
5 x 13 = 65	15 x 13 = 195
6 x 13 = 78	16 x 13 = 208
7 x 13 = 91	17 x 13 = 221
8 x 13 = 104	18 x 13 = 234
9 x 13 = 117	19 x 13 = 247
10 x 13 = 130	20 x 13 = 260

1. Make a subtraction sum so that the answer is a multiple of 13.
(23 - 10 = 13, 30 - 4 = 26)
2. Now change this into years and months (23 years and 10 months)
3. How many months are there in 23 years and 10 months?

This becomes 23 x 12 = 276 and 276 + 10 = 286 months.

Another example is 20 years and 7 months.
This becomes 20 x 12 = 240 and 240 + 7 = 247 months.

When the amount of years minus the number of months is a multiple of 13, the total number of months will always be a multiple of 13.

Another example is 45 years and 6 months.
This becomes 45 x 12 = 540 and 540 + 6 = 546 months.

Yet another example is 68 years and 3 months.
This becomes 68 x 12 = 816 and 816 + 3 = 819 months.

Our final example is 21 years and 8 months.
This becomes 21 x 12 = 252 and 252 + 8 = 260 months.

Can somebody explain why or how this happens? Ain't that peculiar?

Questions

Calculate the total number of months in the following:
1. 22 years and 9 months
2. 24 years and 11 months
3. 40 years and 1 month
4. 18 years and 5 months
5. 19 years and 6 months
6. 15 years and 2 months
7. 44 years and 5 months
8. 55 years and 3 months
9. 85 years and 7 months
10. 58 years and 6 months

How many years and months are in the following examples?
1. 1001 months
2. 845 months
3. 676 months
4. 520 months
5. 546 months
6. 338 months
7. 169 months
8. 325 months
9. 403 months
10. 442 months

LESSON 4: THE NUMBER 37

* $6^2 + 1^2 = 37$
* $19^2 - 18^2 = 37$
* $4^3 - 3^3 = 37$
* $37^4 = 1,874,161$
* $18 + 7 + 4 + 1 + 6 + 1 = 37!$

When I was a child, my family and I would board the number 37 bus for our weekly church services in Peckham, London. I began to think about the number 37 and recognised some interesting occurrences.

Here are the multiples of 37 which are less than 1,000.

37	74	111	148	185	222	259	296	333
370	407	444	481	518	555	592	629	666
703	740	777	814	851	888	925	962	999

Let us look at 3-figured multiples of 37 and look at the figures above them. Look at the column that contains 37 x 4.

1036	1073	1110	1147	1184	1221	1258	1295	1332
1369	1406	1443	1480	1517	1554	1591	1628	1665
1702	1739	1776	1813	1850	1887	1924	1961	1998
2035	2072	2109	2146	2183	2220	2257	2284	2331
2368	2405	2442	2479	2516	2553	2590	2627	2664
2701	2738	2775	2812	2849	2886	2923	2960	2997
3034	3071	3108	3145	3182	3219	3256	3293	3330
3367	3404	3441	3478	3515	3552	3589	3626	3663
3700	3737	3774	3811	3848	3885	3922	3959	3996

Let us look at the column which contains 37 x 7.

259
592
925

As you can see, all 3 digit multiples of 37 are anagrams of other 3-digit multiples of 37! The difference between the numbers in the anagram is always 333, which is 37 x 9.

Can someone explain why or how this pattern exists? Ain't that peculiar?

Multiples of 37 which are above 1,000 (minimum of 4 digits) follow a pattern which is different but arguably more fascinating than the pattern previously mentioned.

Here is a table of the multiples of 37 from 1,000 to 4,000.

1036	1073	1110	1147	1184	1221	1258	1295	1332
1369	1406	1443	1480	1517	1554	1591	1628	1665
1702	1739	1776	1813	1850	1887	1924	1961	1998
2035	2072	2109	2146	2183	2220	2257	2284	2331
2368	2405	2442	2479	2516	2553	2590	2627	2664
2701	2738	2775	2812	2849	2886	2923	2960	2997
3034	3071	3108	3145	3182	3219	3256	3293	3330
3367	3404	3441	3478	3515	3552	3589	3626	3663
3700	3737	3774	3811	3848	3885	3922	3959	3996

Just like with the number 9, the number 37 is embedded in all of its multiples. This is how.
1. Take any 4-digit multiple of 37. We will use 2405.
2. Separate the thousands from the hundreds, tens and units. This now becomes 2 and 405.
3. Change the thousands into a unit and add to the 3 digit number which remains.

2 + 405 = 407
407 = 11 x 37

This is how the number 37 is embedded into the multiples of 37.

Here is another example.

Take 2590
This becomes 2 + 590 = 592
592 = 16 x 37

This is how you can recognise a multiple of 37.

T = Thousands
HTU = Hundreds, tens, units
If T + HTU = 37 x TTHU is a multiple of 37

1517 =
1 + 517 = 518 (i.e. 37 x 14)

3848 =
3 + 848 = 851 (i.e. 37 x 23)

2701 =
2 + 701 = 703 (i.e. 37 x 19)

Ain't that peculiar?

ANSWERS

Lesson 1: Changing Units

1. 120	2. 20	3. 30	4. 40	5. 80
6. 320	7. 140	8. 160	9. 270	10. 20
11. 10	12. 30	13. 100	14. 60	15. 150
16. 90	17. 90	18. 420	19. 300	20. 140

Lesson 2: Years and Months Part One

1. 440 months	2. 121 months
3. 44 months	4. 88 months
5. 242 months	6. 308 months
7. 682 months	8. 22 months
9. 110 months	10. 33 months

1. 16 years and 6 months	2. 21 years and 1 month
3. 17 years and 5 months	4. 35 years and 9 months
5. 3 years and 8 months	6. 40 years and 4 months
7. 30 years and 3 months	8. 10 years and 1 month
9. 46 years and 9 months	10. 13 years and 9 months

Available from www.whenweruled.com

PART FIVE

ABOUT THE AUTHORS

ABOUT THE AUTHORS

Robin Walker 'The Black History Man'

Robin Walker 'The Black History Man' was born in London but has also lived in Jamaica. He attended the London School of Economics and Political Science where he read Economics.

In 1991 and 1992, he studied African World Studies with the brilliant Dr Femi Biko and later with Mr Kenny Bakie. Between 1993 and 1994, he trained as a secondary school teacher at Edge Hill College (linked to the University of Lancaster).

Since 1992 and up to the present period, Robin Walker has lectured in adult education, taught university short courses, and chaired conferences in African World Studies, Egyptology and Black History. The venues have been in Toxteth (Liverpool), Manchester, Leeds, Bradford, Huddersfield, Birmingham, Cambridge, Buckinghamshire and London.

Since 1994 he has taught Economics, Business & Finance, Mathematics, Information Communications Technology, PSHE/Citizenship and also History at various schools in London and Essex.

In 2006 he wrote the seminal *When We Ruled*. It is the most advanced synthesis on Ancient and Mediaeval African history ever written by a single author. It established his reputation as the leading Black History educational service provider.

In 2011 and 2012, he wrote a series of e-book lecture-essays available through Amazon including *The Rise and Fall of Black Wall Street, The Black Musical Tradition, The Mysterious Sciences of the Great Pyramid, Intellectual Life and Legacy of Timbuktu, Black Economic Empowerment: Create Your Own Plan to build Great Wealth, Understanding the Book of the Dead, The Equinox and the Real Story behind Easter, If you want to learn early African history START HERE, West African Contributions to Science and Technology, Ancient Egyptian Contributions to Science and Technology* and finally *African American Contributions to Science and Technology*.

In 2013, he co-authored (with Siaf Millar and Saran Keita) *Everyday Life in an Early West African Empire*.

John Matthews

John Matthews was born and raised in London. He attended Greenwich University where he read Accounting and Finance. He is a personal mathematics tutor with over 20 years of experience. He has taught mathematics at one of London's finest institutes of Higher Education. He has also taught at one of London's most recognised Supplementary Schools.

He enjoys using his abundant gifts to empower and motivate learners to activate their hidden mathematical genius.

Speaking Engagements

Looking for a speaker for your next event?

Robin Walker 'The Black History Man' is dynamic and engaging, both as a speaker and a workshop leader. He brings Black or African history alive, making it relevant for the present generation. You will love his perfect blend of accessibility, engagement, and academic rigour where learning becomes fun. He has a lecture called *The African Origin of Mathematics.*

John Matthews is engaging, dynamic and enthusiastic as a tutor and as a public speaker. He can make mathematics both relevant and fun with his unique delivery. Both adult and child learners have all benefitted from John's special talents. He gives lectures to a variety of audiences.

Adults and child learners will be empowered by the lecture called *The Keys To Unlocking Your Hidden Mathematical Genius* and *Things Your Mathematics Teacher Should Tell You.* Others will be amazed and motivated by his lecture *The African Contribution To Mathematics.*

To book Robin Walker for your next event, send an email to historicalwalker@yahoo.com

To book John Matthews for your next event, send an email to john_matthews60@hotmail.co.uk

INDEX

Printed in Great Britain
by Amazon